Measurement Techniques for Carcinogenic Agents in Workplace Air

Prepared by an expert sub-committee of the Health, Safety and Environment Committee of the Royal Society of Chemistry for the Commission of the European Communities, Directorate-General Employment, Social Affairs and Education, Health and Safety Directorate, Division for Industrial Medicine and Hygiene.

Commission of the European Communities

Measurement Techniques for Carcinogenic Agents in Workplace Air

RC268.6
M42
1989

Publication No. EUR 11897 of the
Commission of the European Communities,
Scientific and Technical Communication Unit,
Directorate-General Telecommunications, Information
Industries and Innovation, Luxembourg

© ECSC-EEC-EAEC, Brussels-Luxembourg, 1989

LEGAL NOTICE
Neither the Commission of the European Communities nor any person acting on behalf of the Commission is responsible for the use which might be made of the following information.

British Library Cataloguing in Publication Data

Measurement techniques for carcinogenic agents in workplace
   air.
   1. Man. Cancer. Pathogens: Chemicals
   I. Royal Society of Chemistry   II. Commission of European
   Communities
   616.99'4071

   ISBN 0-85186-098-2

Photoset in Great Britain by
Rowland Phototypesetting Ltd, Bury St Edmunds, Suffolk
Printed by St Edmundsbury Press Ltd, Bury St Edmunds, Suffolk

Contents

Introduction	1
Ethylene Dibromide	3
1,2-Dibromo-3-chloropropane	7
Dimethyl Carbamyl Chloride	9
Chloromethyl Methyl Ether	11
Bis(chloromethyl) Ether	13
Epichlorohydrin	16
Ethylene Oxide	21
Propanolide	25
1,3-Propane Sultone	27
N-Nitrosodimethylamine	29
Propyleneimine	33
N,N-Dimethylhydrazine	35
2-Nitropropane	38
Acrylonitrile	40
o-Tolidine	43
o-Dianisidine and Salts	46
3,3'-Dichlorobenzidine	49
4,4'-Methylene Bis(2-Chloroaniline)	52
2-Nitronaphthalene	55
5-Nitroacenaphthene	58
Diethyl Sulphate	60
Dimethyl Sulphate	62
Hexamethylphosphoric Triamide	65
Arsenic Trioxide	67
Cadmium Chloride	70
Chromates of Zinc, Calcium and Strontium	72

INTRODUCTION

Preamble

This study was carried out by the Royal Society of Chemistry (RSC) under Contract from the Commission of the European Communities (CEC). The work was overseen by some members of the Society's Health, Safety and Environment Committee (HSEC) and performed largely by the Secretariat of the Committee. The HSEC itself consists of members of the RSC whose affiliations include industry, the trade unions, government and independent consultants.

The thirty-one substances studied were selected by the CEC on the basis of their known or suspected carcinogenic properties. The substances covered in this work do not include those already studied under previous CEC Contracts (see Refs. 1, 2, etc.). The contract requirement was to produce recommended analytical methods for each listed substance when present in workshop air and to provide data on any other methods which had been carried out. In addition, the Committee felt that some additional information might be useful by way of introduction to the analytical methodology and therefore brief résumés have been included covering synonyms, manufacture and uses.

Sources of Information

The basis of these reports is largely authoritative reviews and reliable analytical abstracts. The Committee was of the opinion that, in general, sufficient reliable information was available concerning the carcinogenic substances considered to obviate the necessity of consulting primary sources. However, in certain cases, particularly where the analytical quality assurance data had not been quoted in the review of secondary source, the original papers were consulted. The information has been compiled selectively, with material of doubtful validity or relevance being omitted.

The more familiar and greater used substances have obviously attracted greater study and *vice versa*. Hence some substances command a more detailed coverage than others; a few substances have only one or two listed reputable references and hence these methods have been recommended on the basis of being the only ones in existence at this time.

Style

Each chapter has been laid out to include the CA Registry No., synonyms, manufacture, uses and determination in workshop atmosphere. The determination in workshop atmosphere occupies the largest section within each chapter and is divided into recommended sampling procedure followed by recommended measuring procedure: this is followed by the performance characteristics of the recommended method and also a review of other analytical methods.

References

1. 'Organo-chlorine Solvents: Health Risks to Workers', The Royal Society of Chemistry, London, 1986.
2. 'Solvents in Common Use: Health Risks to Workers', The Royal Society of Chemistry, London, 1988.

Ethylene Dibromide

1. CA REGISTRY NO.
106-93-4

2. SYNONYMS
1,2-dibromoethane

3. MANUFACTURE
Hydrohalogenation of ethylene by an addition reaction.

4. USES
4.1   Soil and fruit fumigant.
4.2   As a component of anti-knock additions to petrol.

5. DETERMINATION IN WORKSHOP ATMOSPHERE
5.1   Recommended Sampling Method
    Technique   : Trapping on a solid adsorber
    Sample   : 7 cm × 4 mm i.d. glass tubes containing 100 mg (front) and activated charcoal separated by a 2 mm urethane plug
    Sample flow   : 0.02 to 0.2 l min^{-1} by personal pump
    Sample volume : 0.1 l (at 0.1 ppm) to 25 l
    Extraction   : 1. Place both portions of adsorber in separate 10 ml flasks
                             2. Add 10 ml of 99:1 benzene:methanol containing 1 μg internal standard such as 1,1,2,2,-tetrachloroethane
                             3. Stand for 60 min with occasional agitation

5.2   Recommended Measuring Method
    Technique   : GC
    Column   : 1.8 m × 4 mm borosilicate glass
    Support phase   : 80–100 mesh Gas Chrom Q
    Liquid phase   : 3% OV-210
    Temperature   : Column 50 °C; injection 175 °C; detector 315 °C
    Detector   : ECD ^{63}Ni
    Carrier gas   : N_2
    Carrier gas flow : 35 ml min^{-1}
    Sample size   : 5 μl

5.3   Performance Characteristics
    Range studied   : 0.2 to 2.4 μg ethylene dibromide
    Bias   : Not significant
    Overall precision : SD = 0.044
    LOD   : 0.01 μg per sample

Interferences : None identified
More detailed information is given in ref. 1.

6. OTHER METHODS

6.1 After evaluating several sorbents for 1,2-dibromoethane, activated charcoal was found to be the most efficient.[2] The adsorbed 1,2-dibromoethane was extracted with benzene–methanol (99:1) and the extract was analysed by GC on a stationary phase of 1.5% OV-17 plus 1–95% OV-210. The column was operated at 75 °C and detection was by electron capture.

6.2 Tenax GC is a popular sorbent for many organic vapours, including 1,2-dibromoethane. A 1.5 cm × 6 mm cartridge of Tenax GC (35–60 mesh) had been used and the 1,2-dibromoethane was thermally desorbed at 270 °C on to a 85 m GC capillary column coated with SE-30.[3] The column (one of four different ones) was temperature programmed from 25 to 240 °C at 4 °C min^{-1}.

6.3 1,2-Dibromoethane was one of 28 organic compounds in ambient air which were trapped on a 15 cm × 0.25 in tube containing Tenax GC (60–80 mesh).[4] Air was passed through at 10–15 l min^{-1} and after thermal desorption, 1,2-dibromoethane was separated on a 50 m GC capillary column coated with SP2100. The column was temperature programmed from −90 °C to 140 °C and detection was by electron capture, the limit being 0.01 ppb.

6.4 Marine air samples were passed at 50–100 ml min^{-1} through a tube containing 80–100 mg Tenax TA.[5] Trapped 1,2-dibromoethane was thermally desorbed at 230 °C into a liquid-nitrogen-cooled coil before being thermally desorbed again at 230 °C on to a 60 m GC capillary column coated with dimethylopolysiloxane. The column was temperature programmed from 20 to 170 °C at 3 °C min^{-1} after an initial hold for 3 min and detection was by ^{63}Ni electron capture.

6.5 A tube containing a line of 1 cm of 3% OV-1, 15 cm of Tenax GC (60–80 mesh) and 8 cm of Davison-type silica gel has been used to trap hazardous substances, including 1,2-dibromoethane.[6] After thermal desorption, compounds were separated on a 2 m × 2 mm glass GC column packed with 1% SP-2100 on Carbopack B. Detection of these compounds was by mass spectrometry (MS) and the limit of detection for 1,2-dibromoethane was 1 µg l^{-1}.

6.6 Tenax GC has also been used to trap volatile halogenated hydrocarbons, including 1,2-dibromoethane, from water after bubbling nitrogen through it.[7] Hydrocarbons were desorbed from the Tenax at 200 °C and separated on a 2 m × 3 mm glass GC column packed with 0.2% Carbowax 1500 on Carbopack-C (80–100 mesh). The column was temperature programmed from 35 to 170 °C at 8 °C min^{-1} after an initial hold of 4 min and detection was by flame ionization.

6.7 The ambient air over Stockholm was analysed for 1,2-dibromoethane, 1,2-dichloroethane and benzene by passing the air through a bed of Poropak Q.[8] The organic compounds were thermally desorbed on to a 2 m × 1.8 mm glass GC column packed with 6% poly-*m*-phenyl ether on Tenax

GC (60–80 mesh). The column was temperature programmed by holding at 40 °C for 1 min and taking to 110 °C at 30 °C min^{-1} and then to 185 °C at 10 °C min^{-1}; detection was by MS.

6.8 Automobile exhaust gases, including 1,2-dibromoethane, were trapped on a 3 m × 3 mm column packed with 15% tritolylphosphate on Chromosorb W (80–100 mesh).[9] The gases were ultimately led through a column of powdered lithium aluminium hydride to remove alkyl nitrites, alkyl nitrates and nitroalkanes and thence on to the 12 cm × 5 mm GC column packed with 3% SE-50 on molecular sieve 5A. After separation, the halogenated compounds were detected by electron capture.

6.9 Gaseous samples were led over 10 ml hexane in a test tube which was stoppered and shaken to dissolve 1,2-dibromoethane.[10] The solution was examined by GC on a 1.5 m × 3 mm o.d. stainless steel column packed with either 5.5% DC-200 on Carbopack B (60–80 mesh) or 11% QF-1 on Gas Chrom Q (80–90 mesh). The temperature of either column was 90 °C and detection was by electron capture.

6.10 Volatile organic compounds, including 1,2-dibromoethane, in air were trapped in a vessel cooled to −150 °C.[11] The volatiles, having been concentrated, were thermally desorbed on to a 50 m capillary GC column coated with OV-1 and the column was temperature programmed from −50 °C to 150 °C at 8 °C min^{-1}, the volatiles being detected by electron capture.

6.11 Volatile halocarbons including 1,2-dibromoethane were similarly trapped and analysed.[12]

6.12 Brominated compounds were monitored near an industrial site and after collection the 1,2-dibromoethane and other compounds were separated on a 45 cm × 2 mm GC column packed with 2% OV-101 on Gas Chrom Q.[13] The column was temperature programmed from 200 to 300 °C at 12 °C min^{-1} and detection was by MS.

6.13 In the development of a new detector system photoionization involving the Lyman α-line has been used for the detection of certain compounds, including 1,2-dibromoethane.[14] The sensitivity of this novel detector approaches that of the electron capture detector. Most methods for the determination of 1,2-dibromoethane in workplace atmospheres rely on GC.

6.14 There is an indirect method for the determination of 1,2-dibromoethane by high performance liquid chromatography after its derivatization with silica-supported silver picrate, but this method seems only to have been applied to the analysis of petrol.[15]

6.15 1,2-Dibromoethane in air was determined after passing 250 ml air through ethanolic potassium hydroxide–potassium chromate solution at 0 °C.[16] The reaction produces bromide by hydrolytic debromination and this is allowed to react with p-rosaniline to yield a yellow compound which is extracted with chloroform and the colour measured at 580 nm. The limit of detection is 1 ppm.

REFERENCES
1. National Institute for Occupational Safety and Health, 'Manual of Analytical Methods,' 3rd ed., 'NIOSH Monitoring Methods' DHEW (NIOSH) Publication No. 1008, US Dept of Health, Education & Welfare, 1984.
2. J. B. Mann et al., J. Environ. Sci. Health, Part B, 1980, **15**, 507.
3. K. J. Krost et al., Anal. Chem., 1982, **54**, 810.
4. B. B. Kebbekus and J. W. Bozzelli, Annu. Meet. Air Pollut. Control Assoc., 1982, **75**, 82–6515.
5. T. Class and K. Ballschmiter, Fresenius' Z. Anal. Chem., 1986, **325**, 1.
6. N. E. Springarn et al., J. Chromatog. Sci., 1982, **20**, 286.
7. E. Stottmeister et al., Acta Hydrochim. Hydrobiol., 1986, **14**, 573.
8. A. Jonsson and S. Berg, J Chromatog., 1980, **190**, 97.
9. T. Kojima and Y. Seo, Bunseki Kagaku, 1976, **25**, 855.
10. S. C. Morris et al., J Chromatog., 1982, **246**, 136.
11. W. A. McClenny et al., Anal. Chem., 1984, **56**, 2947.
12. K. Ballschmiter et al., Fresenius' Z. Anal. Chem., 1986, **323**, 334.
13. E. D. Pellizzari et al., Report EPA 560/6-78/002 (PB-286484), 1978.
14. J. N. Driscoll and F. F. Spaziani, Chem. Mag., 1976, **2**(5), 105.
15. S. T. Colgan et al., Anal. Chem., 1986, **58**, 2366.
16. J. R. Rangaswamy et al., J. Ass. Off. Anal. Chem., 1976, **59**, 1262.

1,2-Dibromo-3-chloropropane

1. CA REGISTRY NO.
96-12-8

2. SYNONYMS
3-Chloro-1,2-dibromopropane; DBCP; dibromochloropropane

3. MANUFACTURE
Prepared by the addition of bromine to allyl chloride.

4. USES
The main, and only main use, is as a soil fumigant for the control of nematodes.

5. DETERMINATION IN WORKSHOP ATMOSPHERE
5.1 Recommended Sampling Method
Technique	: Trapping in Chromosorb, and extraction with hexane
Sampler	: Tube 10 cm × 1 cm i.d.
Adsorbent	: 1 g Chromosorb 101
Sample flow	: 2 l min^{-1} for 3 h
Extraction	: The Chromosorb 101 is shaken for 1 min with 20 ml hexane.

5.2 Recommended Measuring Method
Technique	: GC
Column	: Not quoted
Stationary phase	:
Liquid phase	: 1.55% OV-17 + 1.95% OV-210
Temperature	: 115 °C, isothermal
Detector	: ECD ^{63}Ni
Carrier gas	: N_2
Carrier gas flow	: 70 ml min^{-1}
Sample size	: 2 µl

5.3 Performance Characteristics
Range studied	: 0.07 ppb to 20 ppm
Bias	: None quoted
Overall precision	: None quoted
LOD	: 0.014 ppb
Interferences	: None quoted

More detailed information is given in ref. 1.

6. OTHER METHODS
6.1 A general survey of air pollutants, including DBCP, in outdoor and indoor air of New Jersey has been carried out using GC-MS analysis.[2]

6.2 Experiments have been conducted on the storage stability of DBCP (together with 1,3-dichloropropane) which had been trapped on activated charcoal in air sampling tubes.[3]

6.3 Otherwise, the only other method in the literature involves trapping DBCP from air containing the insecticide Nemagon. The DBCP is trapped in dimethylformamide and analysis is carried out by a.c. polarography.[4,5]

REFERENCES
1. J. B. Mann et al., J. Environ. Sci. Health, Part B, 1980, **15**, 519.
2. E. D. Pellizzari et al., Indoor Air Proc. Int. Conf. Indoor Air Qual. Clim. 3rd, 1984, Vol. 4, p. 227.
3. W. N. Albrecht et al., Bull. Environ. Contam. Toxicol., 1986, **36**, 629.
4. R. M. Novik and N. I. Plyngyu, Tear. Prakt. Polyarogr. Metodov Anal., 1973, 19.
5. R. M. Novik et al., Biosfera Chel. Mater. Vses. Simp. 1st, 1973.

Dimethyl Carbamyl Chloride (DMCC)

1. CA REGISTRY NO.
79-44-7

2. SYNONYMS
Dimethyl carbamic chloride;
(dimethylamino) carbonyl chloride;
N,N-dimethylamino carbonyl chloride;
dimethyl carbamic acid chloride;
N,N-dimethyl carbamic acid chloride;
N,N-dimethyl carbamidoyl chloride;
dimethyl carbamyl chloride;
N,N-dimethyl carbamoyl chloride;
dimethyl chloroformamide;
dimethyl carbamyl chloride.

3. MANUFACTURE
By the reaction of dimethylamine and phosgene.

4. USES
Mainly used as an alkylating agent in the synthesis of herbicides, pesticides and pharmaceuticals.

5. DETERMINATION IN WORKSHOP ATMOSPHERE
5.1 Recommended Sampling Method
Technique	: Adsorption onto solid adsorber
Adsorber	: 80 mg of 60–80 mesh Tenax GC in a 6 in × 0.25 in collection tube
Sample flow	: 0.2 l min^{-1}
Sample volume	: 48 l (0.2 l min^{-1} for 4 h)
Extraction	: Thermal desorption at 300 °C

5.2 Recommended Measuring Method
Technique	: GC
Column	: 6 ft × 0.25 in glass
Support phase	: 80–100 mesh Supelcoport
Liquid phase	: 10% Dexsil 300
Temperature	: column – isothermal at 105 °C for 4 min after sample desorption, then to 165 °C at 8 °C min^{-1}; injector 250 °C; detector 780 °C
Detector	: N$_2$–P
Detector gas	: He at 40 ml min^{-1}; H$_2$ 14 p.s.i.; air 60 p.s.i.
Carrier gas	: He/H$_2$
Carrier gas flow	: He 40 ml min^{-1}; H$_2$ 30 ml min^{-1}

5.3 Performance Characteristics

Range studied	: 0.08 to 1.3 ppb
Bias	: None stated
Overall precision	: Not quoted
LOD	: 10 ppm
Interferences	: None quoted

More detailed information is given in ref. 1.

6. OTHER METHODS

The only other viable method for the determination of DMCC is a colorimetric one.[2] Twenty litres of air containing DMCC is drawn, at a flow-rate of 1 l min^{-1}, through an impinger containing 10 ml of a 1% acetone solution of 4-(p-nitrobenzyl) pyridine. After some manipulation of the trapping solution, its absorbance is measured at 475 nm. The claimed limit of detection is 17 ppb.

REFERENCES

1. L. J. Maienzo and C. J. Hensler, *Am. Ind. Hyg. Assoc. J.*, 1982, **43**, 838.
2. G. M. Rusch *et al.*, *Anal. Chem.*, 1976, **48**, 2259.

Chloromethyl Methyl Ether

1. CA REGISTRY NO.
107-30-2

2. SYNONYMS
CMME; methane, chloromethoxy; chlorodimethyl ether; chloromethoxymethane; α,α-dichlorodimethyl ether; methoxychloromethane; methoxymethylchloride; methyl chloromethyl ether; monochloromethyl methyl ether.

3. MANUFACTURE
Synthesized from aqueous formaldehyde, methanol and hydrogen chloride; anhydrous conditions give rise to the undesirable bis(chloromethyl) ether.

4. USES
4.1 In the manufacture of type I ion exchange resins.
4.2 Undergoes an addition reaction with isoprene to give the 1,2- and 1,4-adducts.

5. DETERMINATION IN WORKSHOP ATMOSPHERE
5.1 Recommended Sampling Method
Technique	: Solid adsorber in 7 cm × 5 mm i.d. glass tube
Adsorber	: 2 g of 1.5% potassium 2,4,6-trichlorophenate on glass beads (120–140) with a back-up tube containing 1 g silica gel
Sample	: Procedure gives a 4 h T.W.A.
Sample flow	: 5–10 ml min^{-1}
Extraction	: The adsorbent is desorbed by shaking with 2 ml methanol for 5 min; the methanol is drawn out into a 3 drm vial – the adsorbent is washed with a further 1 ml methanol. The combined extracts are made alkaline with 3 ml N KOH and extracted with 2 ml hexane – the hexane solution provides the sample for analysis

5.2 Recommended Measuring Method
Technique	: GC
Column	: 7 ft × 2 mm i.d. glass
Packing	: Parabond-PEG20M system
Carrier gas	: N_2
Carrier gas flow	: 20 ml min^{-1}
Temperature	: Oven 140 °C isothermal; detector 320 °C; injector 200 °C
Detector	: ECD
Aliquot	: 1 µl

5.3 Performance Characteristics
  Range studied   : Not given
  Bias   : Not stated
  Overall precision : Not given
  LOD   : Not given

More detailed information is given in ref. 1.

6. OTHER METHODS

6.1 A solution of 0.5% sodium 2,4,6-trichlorophenate was used in two impingers to trap CMME and also bis(chloromethyl) ether (BCME).[2] Air was drawn through the solutions at 500 ml min^{-1} for up to 2 h, and they were afterwards heated, cooled, diluted with water and extracted with hexane. The subsequent GC analysis of the extract was carried out on a 6 ft × 0.25 in glass column packed with 0.1% QF-1 plus 0.1% OV-17 on glass beads (120–200 mesh), operated at 140 °C, detection being by ^{63}Ni ECD, the LOD being 0.02 ng ml^{-1}.

6.2 As little as 2 ml of air containing CMME was injected onto a pre-column of Chromosorb 101 and then passed to a derivatization column (4 in × 0.125 in stainless steel) containing 0.1% OV-275 impregnated with 2,4,6 trichlorophenate supported on glass beads (120–140 mesh).[3] The effluent containing the derivative was subsequently analysed by GC on a 5 ft × 0.125 in stainless steel column packed with 0.1% OV-1 on glass beads (120–140 mesh) operated at 130 °C. Detection was by ECD, the LOD being 16 pg 10 ml^{-1} sample.

6.3 Air containing CMME has been sampled by drawing it through a 3 in × 0.375 in glass tube containing glass beads (60–80 mesh) coated with sodium ethoxide (or other sodium alkoxide).[4] After sampling was completed, nitrogen was used to purge the derivative from the tube onto a GC column for analysis. A similar trapping device, but sodium phenate functioned as the derivatizing agent.[5,6]

6.4 Having trapped the CMME and BCME from the air, various derivatives can be made with either an alkali metal salt of an alcohol, a phenol, a chlorophenol, a chlorinated pyridinol or a chlorinated thiophenol.[7] Any of these derivatives may then be subjected to GC analysis for quantifying the two ethers.

REFERENCES
1. M. L. Langhorst et al., Am. Ind. Hyg. Assoc. J., 1981, **42**, 47.
2. E. Sawicki, Health Lab. Sci., 1976, **13**, 78.
3. G. J. Kallos et al., Anal. Chem., 1977, **49**, 1817.
4. E. F. Ault and R. A. Solomon, Br. Patent 1,479,015.
5. E. F. Ault and R. A. Solomon, Can. Patent 1,040,988.
6. E. F. Ault and R. A. Solomon, Ger. Offen. 2,434,901.
7. E. F. Ault and R. A. Solomon, Ger. Offen. 2,433,614.

Bis(chloromethyl) Ether (BCME)

1. CA REGISTRY NO.
542-88-1

2. SYNONYMS
Methane, oxybischloro; chloromethylether;
dichlorodimethyl ether;
2,2'-dichlorodimethyl ether;
sym-dichlorodimethyl ether;
oxybis[chloromethane].

3. MANUFACTURE
Not produced as such, but occurs as an undesirable byproduct.

4. USES
No specific uses, but occurs as an undesirable byproduct in several manufacturing processes that involve aniline or formaldehyde and hydrochloric acid, *e.g.*, in the manufacture of ion-exchange resins if the acid:aniline ratio exceeds 1.0 then bis(chloromethyl) ether results.

5. DETERMINATION IN WORKSHOP ATMOSPHERE
5.1 Recommended Sampling Method
    Samples : collection tube 12 cm × 2 cm i.d.
    Adsorber : A 9 cm length filled with Tenax GC or Carbopack B
    Sample volume : 10 l at 5–10 ml min^{-1}
    Sample preparation : Thermal desorption at 300 °C directly on to column

5.2 Recommended Measuring Method
    Technique : GC
    Column : 1.5 m × 2 mm i.d. glass
    Support phase : Carbopack C (80–100 mesh)
    Liquid phase : 0.2% Carbowax 20M – this gives the highest resolution; alternatives are 0.1% SP-1000 or 0.1% SP2250
    Temperature : 80–110 °C at 3 °C min^{-1}
    Detector : FID or ECD
    Carrier gas : N_2
    Flow : 20 ml min^{-1}
    Sample size : N/A

5.3 Performance Characteristics
    Range studied : 0.8 to 40 ng BCME 10 l^{-1}
    Bias : None found
    Precision : Overall uncertainty ±25%

LOD	: <0.1 ppb
Interferences	: None reported

More detailed information is given in ref. 1.

6. OTHER METHODS

6.1 Tenax GC has also been used as the absorbent for collecting BCME where GC separation is followed by a chlor-selective flame photometer for determination.[2]

6.2 Tenax GC (60–80 mesh) (175 mg) or Porapak Q (80–100 mesh) (175 mg) has been used to collect BCME.[3] The glass collecting tube was 7.5 cm long with a 5 mm internal diameter and 15 l of air was passed through at a low flow-rate. The BCME was thermally desorbed at 110 °C prior to GC using a dual column system in series with flame ionization detection, the limit of detection being 0.1 ppb.

6.3 Porapak Q has also been used as the collecting material for BCME from air, followed by GC-MS determination.[4,5]

6.4 Trace organic pollutants including BCME in air have been trapped on 200 mg Chromosorb 101 contained in a 3 ft × 0.25 in stainless steel tube.[6] Thermal desorption at 275 °C was followed by GC analysis using flame ionization detection, the limit of detecting being 1 ppb.

6.5 Chromosorb 101 was also used as the trapping medium prior to a GC-MS analysis.[7]

6.6 MS has been favoured by many workers as the detection system following the GC separation of BCME.[8–13]

6.7 BCME in air was determined, after collection, by GC analysis of the derived bis(p-phenyl phenoxymethyl) ether following reaction with sodium p-phenylphenoxide in dimethylformamide.[14] Air was sucked at 0.2–1.2 l min^{-1} through Tenax GC (10–30 l sample) and dimethylformamide was used as extractant before derivatization. The derivative was subjected to GC analysis on a 25 m capillary column coated with SE-54 using flame ionization detection, the limit of detection being 0.1 ppb.

6.8 BCME, after collection from air in hexane, was converted to bis(ethylthiomethyl) ether by reaction with sodium ethanethiolate.[15] The derivative was subjected to GC analysis on a 1 m × 3 mm Teflon column packed with 25% 1,2,3-tris(2-cyanoethoxy)propane on Chromosorb WAW (60–80 mesh). The column was operated at 150 °C and detection was by flame ionization.

6.9 BCME was separated from chloromethyl methyl ether (CMME) after conversion to one of several derivatives.[16] Alkali metal salt of an alcohol, a phenol, a chlorophenol, a bromophenol, chlorinated pyridinol or a chlorinated thiopheol can be reacted with BCME and CMME and any of the derivatives can be separated and analysed by GC.

6.10 A chlorinated phenol derivative was also used in a similar separation.[17]

6.11 A review of the methods for the trace determination of BCME in air has been published.[18]

REFERENCES
1. F. Bruner et al., Anal. Chem., 1978, **50**, 53.
2. D. Ellgehausen, Swiss Chem., 1981, **3**(3A), 95.
3. L. S. Frankel and R. F. Black, Anal. Chem., 1976, **48**, 732.
4. L. Collier, Environ. Sci. Technol., 1972, **6**, 930.
5. G. Muller et al., Ind. Arch. Occup. Environ. Health, 1981, 48, 325.
6. D. G. Parkes et al., J. Am. Ind. Hyg. Assoc., 1976, **37**, 165.
7. L. A. Shadoff et al., Anal. Chem., 1973, **45**, 2341.
8. K. P. Evans et al., Anal. Chem., 1975, **47**, 821.
9. E. Sawicki, Health Lab. Sci., 1976, 13, 78.
10. E. Sawicki, Health Lab. Sci., 1975, **12**, 403.
11. F. L. Schulting and E. R. J. Wils, Anal. Chem., 1977, **49**, 2365.
12. M. Cuba et al., Stud. Cercet. Fiz., 1986, **38**, 678.
13. B. L. Van Duuren et al., (refers to ref. 4: correspondence).
14. L. G. J. Van der Ven and A. Venema, Anal. Chem., 1979, **51**, 1016.
15. Y. Baba and T. Tanaka, Bull. Chem. Soc. Jpn., 1978, **51**, 317.
16. E. F. Ault and R. Solomon, Ger. Offen. 2,433,614.
17. Dow Chemical Co., Neth. Appl. 7,409,880.
18. G. Muller et al., Komb. Bellastungen Arbeitsplatz, Ber. Jahrestag. Dtsch. Ges. Arbeitsmed. 22nd, 1982, p. 553.

Epichlorohydrin

1. CA REGISTRY NO.
106-89-8

2. SYNONYMS
Oxirane (chloromethyl);
propane, 1-chloro-2,3-epoxy;
(chloromethyl) ethylene oxide;
(chloromethyl) oxirane;
2(chloromethyl) oxirane;
chloropropylene oxide;
γ-chloropropylene oxide;
3-chloropropylene oxide;
α-epichlorohydrin;
2,3-epoxypropyl chloride;
glycerol epichlorohydrin;
glycidyl chloride.

3. MANUFACTURE
3.1 Chlorohydrination of allyl chloride with chlorine water to give isomeric glycerol chlorohydrins – alkaline dehydrochlorination and steam stripping give epichlorohydrin.
3.2 Chlorination of acrolein and reduction, with s-butyl alcohol with Al butoxide as catalyst, of the resulting 2,3-dichloropropionaldehyde gives a glycerol β,γ-dichlorohydrin which yields epichlorohydrin upon dehydrochlorination with $Ca(OH)_2$.
3.3 Epoxidation of allyl chloride with peracids or perborates.
3.4 Epoxidation of alkyl chloride with t-butyl hydroperoxide over V, W or Mo compounds.

4. USES
4.1 Bulk of production used in the production of glycerol and epoxy resins.
4.2 Adhesives, anion exchange resins, polymers and paper sizing agents are the main other uses.

5. DETERMINATION IN WORKSHOP ATMOSPHERE
5.1 Recommended Sampling Method

    Technique : Trapping on solid adsorbent
    Sampler : 100 mg 20–40 mesh activated charcoal are placed in a 7 cm × 4 mm glass tube and separated from 50 mg activated charcoal by a 2 mm urethane foam plug
    Sample flow : 0.01 to 0.2 l min^{-1} by personal sampling pump
    Sample volume : 2 l (at 2 ppm) to 30 l

	Extraction	: Place front and rear content section in separate vials, add 1.0 ml CS_2, seal and stand for 30 min with occasional agitation

5.2 Recommended Measuring Method

Technique	: GC
Column	: 1.8 m × 2 mm (i.d.) glass
Support phase	: 80–100 mesh Chromosorb 101
Liquid phase	:
Temperature	: Column 135 °C isothermal; injector 175 °C; detector 215 °C
Detector	: FID
Carrier gas	: N_2 or He
Carrier gas flow	: 2 ml min^{-1}
Sample size	: 5 µl

5.3 Performance Characteristics

Range studied	: 12 to 43 mg m^{-3} (20 l samples)
Bias	: Not significant
Overall precision	: SD = 0.057
LOD	: 0.001 mg per sample
Interferences	: None identified

More detailed information is given in ref. 1.

6. OTHER METHODS

6.1 Epichlorohydrin, together with ethylene glycol, was determined in air samples by the chromatropic acid method using periodic acid as the oxidant.[2] The level of sensitivity was 2 µg in a 1 ml sample. In a similar method, potassium permanganate was used as oxidant.[3]

6.2 In a method for the determination of epichlorohydrin and other substances used in the production of epoxy resins the oxidant was periodic acid after hydrolysis of epichlorohydrin to glycerol.[4] Acetone and triethylene tetramine interfered by forming some formaldehyde, and phenol, if present, reacted with the formaldehyde, which invalidated the method.

6.3 Where formaldehyde already present in the air of industrial buildings was a problem, this was removed by trapping in potassium hydroxide solution before the epichlorohydrin analysis.[5]

6.4 In a more detailed method, epichlorohydrin, together with dichlorohydrin, allyl alcohol and allyl chloride, were determined in the air of a glycerol manufacturing plant.[6] A litre of sample was bubbled through two absorbers, each containing 3 ml of 40% sulphuric acid, to absorb the epichlorohydrin. The two solutions were combined and a 3 ml aliquot was taken and to this was added 0.2 ml of 1.5% of potassium iodate solution in 10% sulphuric acid. After standing for 30 min, sodium sulphate solution and a solution of the sodium salt of chromatropic acid in sulphuric acid were added and after 30 min immersion in a boiling-water bath and dilution with water, the absorbance was measured at 570 nm.

6.5 A similar method was used on 10–20 l air samples, the working range being 5–30 μg of epichlorohydrin.[7]

6.6 Other similar methods have been recorded in the literature.[8–11]

6.7 Pentane-2,4-dione was used in place of chromatropic acid in a method for the determination of epichlorohydrin in air samples.[12] Epichlorohydrin was hydrolysed to give 1-chloropropane-2,3-diol and then to formaldehyde. After reaction with the pentane-2,4-dione the absorbance was measured at 412 nm. The limit of detection was 30 μg m^{-3} of air. From 1976, GC methods largely superseded photometric ones, and most are concerned with the determination of epichlorohydrin in industrial or workshop atmospheres where epichlorohydrin is being produced or is an evident impurity during the production of other chlorinated organic compounds.

6.8 Epichlorohydrin at a concentration of 1 mg m^{-3} was determined by GC on a column containing 15% PFMS-5 on Chromaton N-AW using flame ionization detection.[13]

6.9 In epichlorohydrin production, four chlorinated organic compounds were separated from epichlorohydrin after absorption of air at 200 ml min^{-1} in ethanol in micro absorbers.[14] GC was performed on a column packed with 10% neopentyl glycol succinate on kieselguhr (0.2–0.3 mm) using flame ionization detection.

6.10 Workroom air at 200 ml min^{-1} was monitored for the presence of epichlorohydrin and ethylene chlorohydrin by adsorption on 150 mg Amberlite XAD-7 contained in a tube (5 cm × 4 mm).[15] Dichloromethane was used to desorb the compounds and GC analysis was carried out on a glass column (1.5 m × 4 mm) packed with 0.2% Carbowax 1500 on Carbopack C at 85 °C using flame ionization detection. This method gave 99–100% recoveries for the range 4–400 μg epichlorohydrin. Activated charcoal could also be used as adsorbent with carbon disulphide as extractant, but this method gave slightly less good recoveries, *viz.* 89–93%. After storage of the absorbent following collection, recoveries were 100% for the Amberlite XAD-7 method and 70% for the activated charcoal method.

6.11 Epichlorohydrin and 3-chloropropene were sampled from air and determined by GC on either a polyethylene glycol succinate or a DC-200 stationary phase.[16] The detection limit was 0.8 mg m^{-3} air for epichlorohydrin.

6.12 Epichlorohydrin and methanol were sampled from air by adsorption on to activated charcoal, desorbed with carbon disulphide and determined by GC using a column of Carbowax 20M on Chromosorb W AW-DMCS and flame ionization detection.[17]

6.13 Workplace air was examined for epichlorohydrin using adsorption on to activated charcoal in two tubes, the front one containing 100 mg and the rear one 50 mg.[18] Carbon disulphide was used as extractant and GC of the extract was performed to separate epichlorohydrin from other chlorinated organic compounds on a 1.5 m × 2 mm column packed with Poropak Q (100–200 mesh) using flame ionization detection.

6.14 An early GC reference for the determination of epichlorohydrin used a 3 m × 4 mm GC column packed with 20% polyethyleneglycol 1540 on Diatomite brick (0.25–0.5 mm).[19] The column was temperature programmed from 50 to 130 °C and detection was by conductivity.

6.15 A mixture of epichlorohydrin and its hydrolysis products was separated on a 2 m × 3 mm steel column packed with 25% 5Ph 4E polyphenyl ether on Celite 545.[20] The column was temperature programmed from 100 to 180 °C and detection was by flame ionization.

6.16 Most methods used for the determination of epichlorohydrin in air involve trapping followed by solvent desorption but the absorbent, Tenax GC, has also been used where heat is used to desorb the compound directly on to the GC column.[21] Sampling volumes, desorption temperature and other physical conditions were evaluated for a large number of organic compounds, including epichlorohydrin.

6.17 A Tenax GC cartridge was used to trap epichlorohydrin (and ethylene dibromide). A 1.5 cm × 6 cm bed was used and desorption was achieved at 265–270 °C.[22] Up to four different columns were used with temperature programming to separate a variety of organic compounds, the detection limit for epichlorohydrin being 9.6 ng m^{-3}.

6.18 Methods have also been recorded for the determination of epichlorohydrin in air,[23] in headspace above epoxy resins,[24] in headspace above middle cut alkylglycidyl ethers[25] and in process streams.[26]

6.19 Epoxy resins on coated sheets used for food packaging can contain unreacted epichlorohydrin. Diethyl ether was used to extract this prior to its GC determination on a 1.5 m × 0.3 mm glass column packed with 0.2% Carbowax 1500 on Carbopak C (80–100 mesh).[27] Mass spectrophotometric detection was used and the detection limit was 6 ppb in the packaging.

REFERENCES

1. National Institute for Occupational Safety & Health, 'Manual of Analytical Methods', 3rd ed., 'NIOSH Monitoring Methods', DHEW (NIOSH) Method No. 1010, US Dept. of Health, Education & Welfare, 1984.
2. A. Krynska, *Pr. Cent. Inst. Ochr. Pr.*, 1968, **18**(57), 93.
3. E. Sh. Gronsberg, *Nov. Obl. Prom.-Sanit. Khim.*, 1969, 8.
4. A. Krynska, *Pr. Cent. Inst. Ochr. Pr.*, 1973, **23**(76), 55.
5. N. A. Zavorovskaya and Yu. L. Vitrinskaya, *Nauch. Rab. Inst. Okhr. Vses. Tsent. Sov. Prof. Soyuz.*, 1970, (64) 76.
6. T. G. Lipina and A. A. Belyakov, *Gig. Tr. Prof. Zabol.*, 1975, (5), 49.
7. S. Eminger, *Chem. Listy*, 1976, **70**, 872.
8. G. S. Salyamon, *Gig. Sanit.*, 1969, **34**(4), 64.
9. S. D. Zaugol'nikov *et al.*, *Vestn. Akad. Med. Nauk. SSSR*, 1975, (3), 75.
10. A. P. Fomin, *V. sb. Aktual'n. Vopr. Gigieny i Epidemiol.*, 1976, 117.
11. A. P. Fomin, *Sovrem. Probl. Gigieny i Epidemiol.*, 1983, **17**, 45.
12. S. Eminger and I. Vlacil, *Chem. Prum.*, 1978, **28**, 525.
13. L. S. Kiparisova and V. E. Stepanenko, *Gig. Sanit.*, 1976, (6), 54.
14. D. Tolan *et al.*, *Rev. Chim. (Bucharest)*, 1980, **31**, 596.
15. K. Andersson *et al.*, *Chemosphere*, 1981, **10**(2), 143.

16. A. Yu et al., *Zhongguo Yixue Kexueyuan Xuebao*, 1981, **3**, 209.
17. M. Posniak and A. Krynska, *Pr. Cent. Inst. Ochr. Pr.*, 1983, **33**(119), 219.
18. M. T. Lugari, *Acta Nat. Ateneo Parmense*, 1984, **20**(1), 35.
19. F. F. Muganlinskii et al., *Zh. Anal. Khim.*, 1974, 29, 182.
20. L. V. Gurenko et al., *Prom-st. Ser.: Metody Anal. Kontrolya Kach. Prod. Khim. Prom-st.*, 1980, (5), 3.
21. R. H. Brown and C. J. Purnell, *J. Chromatog.*, 1979, **178**, 79.
22. K. J. Krost et al., *Anal. Chem.*, 1982, **54**, 810.
23. R. S. Kamalov, *Gig. Sanit.*, 1979, (2), 50.
24. J. Widomski and W. Thompson, *Chromatog. Newslet.*, 1979, **7**(2), 31.
25. A. H. Ullman and R. Houston, *J. Chromatog.*, 1981, **211**, 398.
26. S. Husain and P. N. Sarma, *J. Chromatog.*, 1986, **355**, 313.
27. J. B. H. Van Lierop, *J. Chromatog.*, 1978, **166**, 609.

Ethylene Oxide

1. CA REGISTRY NO.
75-21-8

2. SYNONYMS
Oxirane; dihydroxirane;
dimethylene oxide; epoxyethane;
1,2-epoxyethane; ethane oxide;
oxycyclopropane; oxane; oxidoethane; oxyfume.

3. MANUFACTURE
3.1 Direct oxidation of ethylene by oxygen using a silver catalyst.
3.2 Reaction between ethylene and hypochlorous acid to give ethylene chlorohydrin which, in turn, is reacted with calcium or sodium hydroxide (the 'chlorohydrin process').

4. USES
4.1 As an intermediate in the production of ethylene glycol, non-ionic surfactants and ethanolamine.
4.2 As a fumigant and sterilant.

5. DETERMINATION IN WORKSHOP ATMOSPHERE
5.1 Recommended Sampling Method

Technique	: Trapping on solid adsorber
Adsorber	: 45 mm × 4 mm i.d. glass tube containing 100 mg HBr coated charcoal in front part and, separated by a silanized glass wool plug, 50 mg HBr coated charcoal
Sample flow	: 0.05 to 0.15 l min^{-1}
Sample volume	: 1 l (at 5 ppm) to 24 l
Extraction	: 1. Transfer front and rear sorbent portions to separate vials (5 ml vol with PTFE screw caps)
	2. Add 1 ml N,N-dimethylformamide to each
	3. Shake for 10 s, and allow to stand for 5 min
	4. Pipette 20 μl aliquot to a fresh 5 ml vial containing 2.0 ml 2% N-heptafluorobutanoylimidazole in iso-octane
	5. Cap and shake for 1 min – stand at room temp. for not less than 5 min
	6. Add 2 ml pure water – mix for 1 min to ensure complete hydrolysis
	7. Transfer a minimum of 1 ml of iso-octane layer to fresh 2 ml vial

5.2 Recommended Measuring Method

Technique	: GC
Column	: 3 m × 4 mm glass
Support phase	: 80–100 mesh Chromosorb WHP
Liquid phase	: 10% SP-100
Temperature	: Column 100 °C; injector 200 °C; detector 300 °C
Detector	: ECD
Carrier gas	: 95:5/A:CH_4
Carrier gas flow	: 25 ml min^{-1}
Sample size	: 1 µl

5.3 Performance Characteristics

Range studied	: 0.04 to 0.98 ppm (24 l sample)
Bias	: Not significant
Overall precision	: SD = 0.13
LOD	: 1 µg per sample
Interferences	: 2-Bromoethanol

More detailed information is given in ref. 1.

6. OTHER METHODS

6.1 A two-sectioned tube (15 cm × 8 mm) containing JXC activated charcoal was used to trap ethylene oxide from air sampled at 20–100 ml min^{-1} to give a 6–8 l sample.[2] The ethylene oxide was extracted with carbon disulphide and the extract was analysed by GC on a 6.1 m × 3.17 mm stainless steel column packed with 30% Tergitol TMN plus 3% sodium methylate on Chromosorb W, NAW (60–80 mesh). The column was operated at 60 °C and detection was by flame ionization, the limit of detection being 0.15 ppm. There is also an alternative GC method with similar performance characteristics.

6.2 In a similar method, a double glass tube containing 800 mg activated charcoal in the front portion and 200 mg in the rear portion was used to trap ethylene oxide from air.[3] The sampling rate was 500 ml min^{-1} for an 8 h period and the ethylene oxide was ultimately extracted with N,N-dimethylacetamide. The extract was analysed on a 2 m × 0.125 in column coated with 15%FFAP on Anakrom A (100–110 mesh) operated at 70 °C. Detection was by flame ionization, the limit of detection being 0.1 ppm.

6.3 According to some workers, activated charcoal as a trapping medium for ethylene oxide has some shortcomings, *i.e.*, the marked effects of relative humidity and ambient temperature.[4]

6.4 Instead, activated charcoal treated with hydrobromic acid was preferred whereby the ethylene oxide reacts to give 2-bromoethanol.[5] Twenty litres of air was sampled at 500 ml min^{-1} through 1 g of the treated charcoal and the 2-bromoethanol was extracted with methanoldichloromethane (9:1). The extract was analysed by GC on a 4 m × 0.125 in stainless steel column

6.5 Hydrobromic acid-coated activate charcoal was used in two tubes of 100 mg (front) and 50 mg (rear) to trap ethylene oxide from 24 l of air sampled at 100 ml min^{-1}.[6] Dimethylformamide was used to extract the 2-bromoethanol formed and was transferred to a vial containing n-heptafluorobutyrylimidazole in iso-octane. After reaction and dilution, the derivative was chromatographed on a 6 ft × 2 mm glass column packed with 10% SP-1000 on Supelcoport (80–100 mesh) operated at 85 °C. Detection was by electron capture, the limit of detection being 0.29 pg.

6.6 Ambersorb XE347, treated with hydrobromic acid, has also been used for trapping ethylene oxide.[7] A 5 cm × 5 mm tube containing 500 mg was used, sampling air at 20 ml min^{-1} for a period of 4–8 h. Acetonitrile–toluene (1:1) containing sodium carbonate was used to extract the 2-bromoethanol and the extract was subjected to GC on a 12 ft × 0.125 in stainless steel column packed with diethylene glycol succinate on Chromosorb W HP (80–100 mesh). The column was operated at 155 °C and detection was by electron capture, the limit of detection being 0.5 ppm.

6.7 Air containing ethylene oxide was drawn through an impinger containing 0.1N sulphuric acid on glass beads at a flow-rate of 6–8 l h^{-1} to give a sample of 4 l.[8] After neutralization, the solution containing the resulting ethylene glycol was chromatographed on a 1.2 m × 4 mm glass column packed with 5% Igepol Co-990 on Teflon T-6 (40–60 mesh). The column was operated at 160 °C and detection was by flame ionization.

6.8 Direct injection of ethylene oxide was made (1 ml) on to a GC column packed with 30% dodecyl phthalate on Chromosorb W (40–60 mesh).[9] The column was operated at 50 °C, the feature of the method being the use of a thermistor as a detector. The working range of the method is 1–1000 mg l^{-1}.

6.9 Before GC became the method of choice, ethylene oxide was adsorbed from air at 500 ml min^{-1} on 2 g of silica gel (40–60 mesh) contained in a 18 cm × 6 mm tube.[10] The silica gel was extracted with water, 0.2M periodic acid was added and the mixture heated for 40 min to form formaldehyde. The solution was then reacted with 0.5M sodium arsenite in acetyl acetone and the resulting colour was measured at 412 nm.

6.10 A more recent semi-quantitative method has used the intensity of a stain produced by the reaction of a quinoidal compound with ethylene oxide.[11]

REFERENCES

1. National Institute for Occupational Safety & Health, 'Manual of Analytical Methods', 3rd ed., 'NIOSH Monitoring Methods', DHEW (NIOSH) Publication No. 1607, US Dept. of Health, Education & Welfare, 1984.
2. A. H. Quazi and N. H. Ketcham, *Am. Ind. Hyg. Assoc. J.*, 1977, **38**, 635.
3. R. Binetti *et al.*, *Chromatographia*, 1986, **21**, 701.
4. G. Graff *et al.*, *Chromatographia*, 1986, **21**, 201.
5. C. Lefevre *et al.*, *Chromatographia*, 1986, **21**, 269.

6. S. P. Tucker and J. E. Arnold, Report 1984, Order No. PB84-242049.
7. G. C. Esposito *et al.*, *Anal. Chem.*, 1984, **56**, 1950.
8. S. J. Romano and J. A. Renner, *Am. Ind. Hyg. Assoc. J.*, 1979, **40**, 742.
9. T. Dumas, *J. Chromatog.*, 1976, **121**, 147.
10. W. A. M. Dan Tonkelaar, *Atmos. Environ.*, 1969, **3**, 481.
11. I. M. Pritts *et al.*, *ASTM Spec. Tech. Publ.*, 1982, **768**, 14.

Propanolide

1. CA REGISTRY NO.
57-57-8

2. SYNONYMS
2-Oxetanone; hydracrylic acid;
β-lactone; 3-propanolide;
propiolactone; BPL;
β-propiolactone; 1,3-propiolactone;
3-propiolactone; β-propionolactone.

3. MANUFACTURE
Addition of a ketene to a carbonyl bond.

4. USES
As an intermediate in the production of acrylic esters.

5. DETERMINATION IN WORKSHOP ATMOSPHERE
5.1 Recommended Sampling Method
Technique	: Collection and concentration on solid adsorber
Adsorber	: 60–80 mesh Tenax GC in sampling tube
Sample flow	: < 1 l min^{-1}
Sample volume	: Not stated
Extraction	: Thermal desorption at 300 °C directly into GC column

5.2 Recommended Measuring Method
Technique	: GC
Column	: 2 m × 0.25 in glass
Support phase	: Chromosorb W
Liquid phase	: Carbowax 20M
Temperature	: Column −30 °C to 120 °C at 30 °C min^{-1}
Detector	: FID
Carrier gas	: N_2
Carrier gas flow	: Not stated
Sample size	: Whole output from thermal desorber

5.3 Performance Characteristics
Range studied	: Not given
Bias	: Not stated
Overall precision	: Not stated
LOD	: 0.1 ppb
Interferences	: None given

More detailed information is given in ref. 1.

6. OTHER METHODS
6.1 In a similar method a 6 cm × 1.5 cm bed of Tenax GC (35–60 mesh) was used

to adsorb propanolide from air.[2] The propanolide was desorbed at 270 °C on to one of four columns, e.g., a 85 m capillar column coated with SE-30. This column was temperature programmed from 25 to 240 °C at 4 °C min^{-1} and detection was by mass spectrometry. The limit of detection was 3 ng m^{-3} air.

6.2 Propanolide has been determined in soil extracts by GC of the bromophenacylate derivative using mass spectrometric detection.[3]

6.3 In the development of a personal badge for monitoring propanolide in air, cellulose strips treated with 4-(*p*-nitrobenzyl)pyridine and deoxyguanosine were successful, a blue colour resulting from the reaction.[4]

REFERENCES
1. A. Agostiano *et al.*, *Water, Air, Soil, Pollut.*, 1983, **19**, 309.
2. K. J. Krost *et al.*, *Anal. Chem.*, 1982, **54**, 810.
3. K. Matsunaja *et al.*, *Okayama-Ken Kankyo Hoken Senta Nenpo*, 1985, **9**, 170.
4. A. Segal and G. Lowewengant, *Mater. Am. Chem. Soc. Meet.*, 1979, **1**, 197.

1,3-Propane Sultone

1. CA REGISTRY NO.
1120-71-4

2. SYNONYMS
1,2-Oxathiolane-2,2-dioxide;
1-propanesulphuric acid;
3-hydroxy-γ-sultone; propane sultone; γ-propane sultone.

3. MANUFACTURE
By the dehydration of γ-hydroxypropane sulphonic acid, *via* sodium hydroxypropane sulphonate from allyl alcohol.

4. USES
4.1 As an intermediate to introduce the sulphopropyl group.
4.2 To confer water solubility and an anionic character.

5. DETERMINATION IN WORKSHOP ATMOSPHERE
5.1 Recommended Sampling Method
Technique	: Adsorption onto the walls of a separation tube.
Adsorber	: A glass separation (stripper) tube, 75 cm × 6 mm i.d., having a thin coating, on the inside, of 2-mercaptobenzothiazole
Sample flow	: 330 ml min^{-1}
Sample volume	: Flow maintained for a time dependent on concentration
Extraction	: The stripper tube was extracted with 100 μl of double distilled water

5.2 Recommended Measuring Method
Technique	: HPLC
Column	: 5 cm × 5 mm, WGA
Packing	: 5 μm ODS (RP-18) – Hypersil
Mobile phase	: 40:60 methanol:0.05M aq. NaHCO$_3$
Temperature	: Ambient
Detector	: U.v. at 286 nm
Carrier gas flow	: 1.5 ml min^{-1}
Sample size	: 10 μl

5.3 Performance Characteristics
Range studied	: 1 to 16 μg
Bias	: None recorded
Overall precision	: RSD = 6% (at 4 μg)
LOD	: 0.5 μg
Interferences	: None recorded

More detailed information is given in ref. 1.

6. OTHER METHODS

6.1 The same workers have produced a gas chromatographic (GC) method following an impinger trapping method.[1] Air is drawn through the impinger which contains methyl isobutyl ketone kept cool in a jacket of ice-water. The flow-rate is 330 ml min^{-1} and the air is sampled for 8 h.

GC is performed on a 120 cm × 3 mm glass column packed with 3% OV-25 on Chromosorb WHP (100–120 mesh). The column is temperature programmed from 147 to 207 °C at 12 °C min^{-1} and detection is either by flame photometry or flame ionization. The limit of detection is 1.4 ng.

If sensitivity was of paramount importance, this GC method would be recommended but the sampling involving a liquid would be less easy to manage if personal sampling was made.

REFERENCES

1. J. Oldeweme and D. Klockow, *Fresenius' Z. Anal. Chem.*, 1986, **325**, 57.

N-Nitrosodimethylamine (DMN)

1. CA REGISTRY NO.
62-75-9

2. SYNONYMS
N,N'-Dimethylnitrosamine;
N-methyl-*N*-nitroso methanamine;
dimethylnitrosamine;
dimethylamine, *N*-nitroso; DMN; DMNA; NDMA

3. MANUFACTURE
3.1 Reaction between ammonia and an alcohol over a dehydration catalyst, *e.g.*, Al_2O_3, TiO_2.
3.2 Reaction between ammonia and an alcohol over a dehydrogenation catalyst, *e.g.*, Ag, Ni or Cu.
3.3 Ammonia reacted with an aldehyde or ketone and hydrogen over a hydrogenation catalyst, *e.g.*, Ag, Ni or Cu.
3.4 Reaction of sodium nitrite with an acidified solution of dimethylamine hydrochloride.

4. USES
As an intermediate in the production of 1,1'-dimethylhydrazine, a liquid rocket fuel.

5. DETERMINATION IN WORKSHOP ATMOSPHERE
5.1 Recommended Sampling Method
Sampling	: Trapping and concentration on solid adsorber
Adsorber	: 7 cm × 4 mm i.d. stainless steel tubing containing two portions of 35–60 mesh Tenax GC: 100 mg of adsorber in the front section separated by a 2 mm plug of urethane foam from 50 mg in the rear section
Sample flow	: 0.01 to 0.12 l min^{-1}
Sample volume	: 5 l
Extraction	: Thermal desorption directly into the GC: conditions are: temperature 200 °C; equalization time 1 min; desorbing rate 70 ml min^{-1}, sample time 5 s; gas He **Note:** some thermal desorbers allow an aliquot to be taken.

5.2 Recommended Measuring Method
Technique	: GC
Column	: 10 ft × 0.125 in stainless steel
Support phase	: 0–100 mesh Supelcoport
Liquid phase	: 20% SP-1000

	Temperature	: Column 150 °C; injector 200 °C; detector 200 °C
	Detector	: N–P
	Carrier gas	: He
	Carrier gas flow	: 70 ml min^{-1}
	Sample size	: N/A
5.3	**Performance Characteristics**	
	Range studied	: 2 to 20 µg m^{-3} on a 5 l sample
	Bias	: None noted
	Overall precision	: RSD = 0.037 at 20 µg m^{-3}; 0.129 at 2 µg m^{-3}
	LOD	: 2g m^{-3} using an aliquot of sample of 1/60 (see Note above under 'Extraction')
	Interferences	: Compounds having the same retention time would interfere; however with the selectivity of the detector, such compounds that do *not* contain N$_2$ or P do not interfere.

More detailed information is given in ref. 1.

6. OTHER METHODS

6.1 The trapping of DMN on Tenax GC has been used by many workers, often followed by GC separation and mass spectrometric detection.[2-6]

6.2 Tenax GC has also been used to trap DMN prior to GC with a thermal energy analyser detection system.[7] A bed of Thermosorb N (35–60 mesh) (a mixture of magnesium silicate, an amine-complexing agent and a nitrosating inhibitor) was used to trap DMN from 100 l air samples passed at 2 l min^{-1}.[8] Acetone was used to extract the DMN prior to GC analysis on a 5 m × 3.2 mm column packed with 5% Carbowax 20M on Chromosorb W (100–120 mesh) at 160 °C. Detection was by thermal energy analyser and the limit of detection was 30 ng m^{-3}.

6.3 Air in a tyre storage area was monitored for the presence of DMN by passing it at 2 l min^{-1} for 150 min through Thermosorb N.[9] The DMN was extracted with dichloromethane–methanol (3:1) and the extract was separated by GC on a 50 m capillary column coated with Carbowax 20M. The column was temperature programmed by holding at 50 °C for 2 min followed by 4 °C min^{-1} rise up to 230 °C. Detection was by ms or by thermal energy analyser and the limit of detection was in the range 0.1–0.3 µg m^{-3}.

6.4 Thermosorb N was also used as the trapping medium for DMN which was subsequently determined by GC.[10,11]

6.5 Charcoal has rarely been used as a trapping medium for DMN but a successful method incorporated a 450 mg trap with diethyl ether as the extracting solvent.[12] The subsequent GC analysis was performed on a column of adipic acid or ethylene glycol on Porolith.

6.6 Caustic impinger traps at ambient temperature and also cryogenic traps have been used to collect DMN and nitrous oxide fume at 2.8–4 l min^{-1} from traffic pollution.[13] Dichloromethane was used to extract the potassium hydroxide trapping solutions and the extract was subjected to GC on a 5 ft × 0.25 in stainless steel column packed with 10% Carbowax 20M on Chromosorb W (80–100 mesh). The column was operated at 140 °C and the choice of detection was between flame ionization, ms and thermal energy analyser.

6.7 Potassium hydroxide impingers have also been used to trap DMN at 2 l min^{-1} from air over Paris.[14] Dichloromethane was used to extract the DMN and after evaporation the residue was dissolved in n-hexane for a similar GC analysis to that of the previous method, using thermal energy analyser detection. The limit of detection was 0.3 ng m^{-3}. Potassium hydroxide trapping of DMN has also been used by other workers.[11]

6.8 Cold vessels at −150 °C have been used to trap DMN and other N-containing compounds with the subsequent GC analysis separating DMN from nitrate and nitrite.[15] The effluent was pyrolysed at 300–500 °C and the NO produced was reacted with ozone and the resulting chemiluminescence was measured.

6.9 Chemiluminescence resulting from the same reaction was used to determine DMN concentration after the reaction between effluent DMN and u.v. light to form NO.[16]

6.10 Traps of N potassium hydroxide solution held at either −79 °C or −95 °C have been used to trap DMN.[17] The solution was then extracted with dichloromethane and the extract was analysed by GC on a 6.5 m × 2 mm stainless steel column packed with 15% FFAP on Chromosorb W (80–100 mesh). Detection was by thermal energy analyser and the limit of detection was 1 ng m^{-3}. The thermal energy analyser has become popular for detecting nitrogen containing compounds and this system has been widely used by other workers.[18,19]

6.11 Air has also been examined for DMN and other nitrosamines by GC with flame ionization detection[20] and with MS detection.[21] An MS detection method has also been used where the GC effluent was ionized by a laser.[22]

6.12 Although GC methods have dominated in the determination of DMN, high performance liquid chromatography (HPLC) methods have also been used. Cold trapping of DMN in N potassium hydroxide solution[17] was used prior to dichloromethane extraction and HPLC of the extract.[23] This was performed on a 30 cm × 4 mm column packed with μBondapak CN using 1.5% acetonitrile in iso-octane as mobile phase at 1.5 ml min^{-1}.

6.13 In another HPLC method, the column had the same dimensions but was packed with μBondapak C_{18} and the mobile phase was 5% aqueous acetic acid at 2 ml min^{-1};[24] detection was by thermal energy analyser.

6.14 The same column material was used in a method using methanol–water (4:1) as mobile phase at 1.5 ml min.$^{-1}$[25] In this case, detection was achieved by photoionization conductivity (a combination of u.v. photolysis and electric conductance) and the limit of detection was 50 ng.

6.15 Air containing nitrosodialkylamines, including DMN, was trapped in water which was concentrated by distillation.[26] The nitrosamines were reduced by zinc in hydrochloric acid and subsequently oxidized to monoalkylhydrazines. These compounds were coupled with βdimethylaminobenzaldehyde to give alkylaldazine complexes, the absorbance of these being determined at 458 nm. The limit of detection for DMN was 0.1 µg.

REFERENCES
1. National Institute for Occupational Safety & Health, 'Manual of Analytical Methods', 2nd ed., 'NIOSH Monitoring Methods', DHEW (NIOSH) Publication No. 299, (Vol. 5), US Dept. of Health, Education & Welfare, 1979.
2. E. D. Pellizzari et al., *Anal. Lett.* 1976, **9**, 579.
3. E. Sawicki, *Health Lab. Sci.*, 1977, **14**(1), 59.
4. J. T. Bursey et al., *Int. Lab.*, 1978, Jan–Feb, 11.
5. J. T. Bursey et al., *Am. Lab. (Fairfield Conn.)*, 1977, **9**(13), 35.
6. J. P. Conkle et al., *Proc. Conf. Environ. Chem. Hydrazine Fuels*, 1977, 63.
7. T. Matamura et al., *Nippon Kagaku Kaishi*, 1979, (10), 1410.
8. D. P. Rounbehler et al., *ASTM Spec. Tech. Publ.*, 1980, **721**, 80.
9. R. S. Morano et al., *Anal. Chem.*, 1982, **54**, 1947.
10. S. Kusumoto et al., *Osaka-furitsu Koshu Eisei Kenkyusho Kenkyu Hokoku, Kogai Eisei Hen 2*, 1981, 1.
11. J. M. Fajen et al., *IARC Sci. Publ.*, 1980, **31**, 517.
12. K. Bretschneider and J. Matz, *IARC Sci. Publ.*, 1976, **14**, 97.
13. R. L. Fisher et al., *Anal. Chem.*, 1977, **49**, 1821.
14. B. T. Choung and M. Benarie, *Stud. Environ. Sci.*, 1980, **8**, 397.
15. D. H. Fine and D. P. Rounbehler, US Patent: 3,996,004.
16. B. Budevska et al., *J. Chromatog.*, 1986, **351**, 501.
17. D. H. Fine et al., *Environ. Sci. Technol.*, 1977, **11**, 577.
18. D. H. Fine et al., *IARC Sci. Publ.*, 1984, **57**, 121.
19. M. Borwitzky, *Chromatographia*, 1986, **22**(1–6), 65.
20. A. Audere et al., *Gig. Tr. Prof. Zabol.*, 1981, (1), 47.
21. C. V. Cooper, *Am. Ind. Hyg. Assoc. J.*, 1987, **48**, 265.
22. R. B. Opsal and J. P. Reilly, *Anal. Chem.*, 1986, **58**, 2919.
23. D. H. Fine et al., *Int. Conf. Environ. Sensing Assess.*, 1975, 2(58), 5pp.
24. J. K. Baker and Cheng-Yu Ma, *IARC Sci. Publ.*, 1978, **19**, 19.
25. B. Bush, et al., *Anal. Lett.*, 1984, 17A, 467.
26. L. Ceh and F. Ender, *Food Cosmet. Toxicol.*, 1978, **16**, 117.

Propyleneimine

1. CA REGISTRY NO.
75-55-8

2. SYNONYMS
2-Methylaziridine; aziridine, 2-methyl;
2-methylethylenimine; 1,2-propylenimine.

3. MANUFACTURE
3.1 By the reaction between ethanolamine and sulphuric acid, followed by ring closure using sodium hydroxide.
3.2 By the reaction between 1,2-dichloropropane with excess ammonia and an inorganic acid acceptor (*e.g.*, CaO) at elevated temperature.

4. USES
Apparently used exclusively as an intermediate in the modification of latex surface coating resins to improve adhesion.

5. DETERMINATION IN WORKSHOP ATMOSPHERE
5.1 Recommended Sampling Method
Technique	: Trapping on solid adsorbent
Adsorber	: Tenax GC (see Note)
Sample flow	: Not $> 1 \, l \, min^{-1}$ (see Note)
Sample volume	: (see Note)
Extraction	: Thermal desorption

Note: No attention has been paid in the literature to the determination of propylenimine in air so Tenax GC is a suggestion – other possibilities are silica gel, activated charcoal or an impinger containing a suitable solvent. In any case conditions of sampling will have to be established.

5.2 Recommended Measuring Method
Technique	: HPLC
Column	: 33 cm × 6 mm
Active phase	: Dowex 50 – Ni
Mobile phase	: 1M Ammonia
Flow	: 38 ml h^{-1} at 15–50 pounds
Detector	: Differential refractometer
Sample size	: 10 mg

5.3 Performance Characteristics
Range studied	: Not given
Bias	: Not given
Overall precision	: Not given
LOD	: 0.01 mg
Interferences	: Not given

More detailed information is given in ref. 1.

6. OTHER METHODS

6.1 In a method for the analysis of aliphatic amines and imines, gas chromatography (GC) was used.[2] The 2 m × 4 mm stainless steel column was packed with 29% UCON LB-550-X plus 20% potassium hydroxide on Chromosorb P and it was operated at 80 °C, detection being by hot wire.

6.2 Another GC method, specifically for ethyleneimine but which might be suitable for propylenimine, was performed on a 18 ft × 0.25 in column packed with 15% Tergitol 35 on Chromosorb W (60–80 mesh).[3] The column was operated at 60–70 °C and detection was by flame ionization.

It is clear that if the determination of propylenimine in workshop atmosphere is to be carried out, research into a suitable sampling technique is urgently required.

REFERENCES

1. K. Shimomura *et al.*, *Anal. Chem.*, 1973, **45**, 501.
2. A. D. Lorenzo and G. Russo, *J. Gas Chromatog.*, 1968, **6**, 509.
3. Encyclopedia of Industrial Chemical Analysis, Vol. 14, p. 478.

N,N-Dimethylhydrazine (DMH)

1. CA REGISTRY NO.
57-14-7

2. SYNONYMS
Hydrazine, 1,1-dimethyl; dimazin;
dimazine; *as*-dimethyl-hydrazine;
u-dimethylhydrazine; UDMM
unsym-dimethylhydrazine;
1,1-dimethylhydrazine.

3. MANUFACTURE
3.1 A modified Raschig reaction where chloramine reacts with dimethylamide.
3.2 Nitrosation and catalytic reduction of dimethylamine – this process has largely been abandoned.
3.3 Reductive catalytic alkylation of a hydrazide (usually acetic acid hydrazide) with formaldehyde and hydrogen followed by basic hydrolysis to remove the acetyl group.

4. USES
As a propellant.

5. DETERMINATION IN WORKSHOP ATMOSPHERE
5.1 Recommended Sampling Method

Technique	: Collection and concentration on solid absorber
Adsorbent	: 400 mg silica gel D-08 (45–60 mesh) in a 8 cm × 6 mm collection tube; the silica gel is in two 200 mg sections separated by a glass wool plug
Sample flow	: 200 ml min^{-1} by means of small pump
Sample volume	: 100 l
Extraction	: Transfer each 200 mg portion of separate 5 ml test tubes, add 2 ml H_2O and stand for 1 h with intermittent shaking; add 2 ml 4% 2-furaldehyde in 0.5 M sodium acetate solution – after 1 h add 0.5 ml ethyl acetate and shake for 1 min. Allow ethyl acetate layer to separate and use aliquot of the ethyl acetate solution for GC analysis.

5.2 Recommended Measuring Method

Technique	: GC
Column	: 1 m × 2 mm glass
Support phase	: 80–100 mesh Supelcoport (or equivalent)
Liquid phase	: 10% OV-7
Temperature	: Column 80 °C for 12 min then to 185 °C at 24 °C min^{-1}, hold for 8 min; injector 150 °C; detector 200 °C

Detector	: FID
Detector gas	: H_2 and air
Detector gas flow	: H_2 at 40 ml min^{-1}
	Air at 540 ml min^{-1}
Carrier gas	: He
Carrier gas flow	: 50 ml min^{-1}
Sample size	: 3 μl aliquots

5.3 Performance Characteristics

Range studied	: 0.04 to 120 mg m^{-3}
Bias	: None stated
Overall precision	: RSD of 0.04 at 1.6 and 3.8 mg m^{-3}
LOD	: None quoted
Interferences	: Compounds having the same retention time as one of the hydrazine derivatives interfere

More detailed information is given in ref. 1.

6. OTHER METHODS

6.1 Tenax GC has been used as an adsorbent for many organic pollutants, including DMH, in air.[2] A stainless steel tube 7.5 cm × 4.5 mm containing 130 mg of Tenax GC (40–60 mesh) was used for air sampled at flow-rates between 5 and 600 ml min^{-1}.

6.2 Acetone has been used to trap DMH, hydrazine and methylhydrazine from air which was passed at 0.5–1.5 ml min^{-1}.[3] The acetone traps the hydrazine derivatives and the resulting hydrazones were extracted with dichloromethane and the extract was analysed by capillary GC on a 60 m column coated with OV-101. The column was operated at 250 °C and detection was by either flame ionization or by thermionic nitrogen–phosphorus detection; the limit of detection was 0.5 μg DMH ml^{-1}.

6.3 Chilled acetone, acidified with acetic acid, was used similarly to trap hydrazines, including DMH, from 2 l of air sampled at 200 ml min^{-1}.[4] The resulting derivatives were chromatographed on a 6 ft × 0.125 in glass column packed with 4% Carbowax 20M plus 0.8% potassium hydroxide on Carbopack B, the column being temperature programmed starting at 110 °C for 4 min and then to 210 °C at 32 °C min^{-1}. Detection was either by thermal energy analyser or by thermionic nitrogen–phosphorus detector, the limit of detection for DMH being given as 0.13 nmol ml^{-1}.

6.4 GC is a preferred technique for the determination of DMH in air, other methods lacking the necessary specificity. These methods include an electrochemical carbon monoxide analyser[5] and the use of an electrochemical cell coupled to an air-sampler system.[6] There is a semi-quantitative method for the determination of missile fuels, including DMH, in the atmosphere whereby air is sucked over crystals of silica gel pre-treated with chromium trioxide and metaphosphoric acid, the resulting colour being measured.[7]

6.5 A novel method has been reported whereby air is passed through a detection system comprising alumina pellets containing approximately 30% of iridium.[8] The DMH and other hydrazines are decomposed with the evolution of heat which is measured by a thermally-responsive thermistor.

REFERENCES
1. National Institute for Occupational Safety & Health, 'Manual of Analytical Methods', 2nd ed., Vol. I, 'NIOSH Monitoring Methods', DHEW (NIOSH) Publication No. 248, US Dept. of Health, Education & Welfare, 1977.
2. R. H. Brown and C. J. Purnell, *J. Chromatog.*, 1979, **178**, 79.
3. J. T. Veal, Proc. Conf. Environ. Chem. Hydrazine Fuels, 1977, p. 79.
4. J. R. Holtzclaw *et al.*, *Anal. Chem.*, 1984, **56**, 2952.
5. L. J. Luskus and H. J. Kilian, *Anal. Lett.*, 1976, **9**, 929.
6. J. R. Stetter *et al.*, *Talanta*, 1979, **26**, 799.
7. Mine Safety Appliances Co., Br. Patent 1,151,594; 7.5.69.
8. E. F. Groomes and J. A. Murfree, US Patent 4,200,608; 26.4.80.

2-Nitropropane

1. CA REGISTRY NO.
79-46-9

2. SYNONYMS
Dimethylnitromethane

3. MANUFACTURE
3.1 Oxidation of amines with n-chloroperbenzoic acid (usually on laboratory scale only).
3.2 Vapour phase nitration of propane and separation from the reaction products (mainly aldehydes and ketones) by washing and fractionation.

4. USES
4.1 As an intermediate in the manufacture of chloropicric acid and nitroacetic acid and a number of other organic chemicals, *e.g.*, hydroxylamine (among others).
4.2 As a solvent, especially in separation processes.

5. DETERMINATION IN WORKSHOP ATMOSPHERE
5.1 Recommended Sampling Method

Technique	: Trapping and concentration on solid adsorber
Adsorber	: 7.0 cm × 4 mm (i.d.) glass tube containing two sections of 60–80 mesh pre-extracted Chromosorb 106, 100 mg in front section separated by a 2 mm urethane foam plug from 50 mg sorbent in rear section
Sample flow	: 0.01 to 0.05 l min^{-1}
Sample volume	: 0.1 to 2 l
Extraction	: 1. Place each section of adsorber in separate vials
	2. Pipette 1 ml ethyl acetate into each vial, cap, and stand for 30 min with occasional agitation

5.2 Recommended Measuring Method

Technique	: GC
Column	: 6 m × 4 mm stainless steel
Support phase	: 80–100 mesh Chromosorb WHP
Liquid phase	: 10% FFAP
Temperature	: Column 90 °C; injector 200 °C; detector 190 °C
Detector	: FID
Carrier gas	: He or N_2
Carrier gas flow	: 20 ml min^{-1}
Sample size	: 5 µl

5.3 Performance Characteristics
Range studied : 3.1 to 28.3 mg m^{-3}
Bias : Not significant
Overall precision : SD = 0.05
LOD : 0.001 mg per sample
Interferences : None found

More detailed information is given in ref. 1.

6. OTHER METHODS
6.1 The NIOSH method is based on the work of Glaser and Woodfin who have published an almost identical method elsewhere.[2]

6.2 There has been very little interest in the determination of 2-nitropropane in air, only one other relevant reference being found in the literature.[3] This method involved trapping air containing aliphatic compounds on various sorbents in order to discover the best adsorbing material. 2-Nitropropane was trapped on 150 mg Amberlite XAD-7 (20–50 mesh) packed into a glass tube (50 mm × 4 mm), which was connected to a personal sampler. Air was passed through the adsorbent at 150 ml min^{-1} to obtain a 5 l sample and 5% methanol in diethyl ether was used to extract the 2-nitropropane. The extract was analysed by GC on a 1.5 mm × 4 mm glass column packed with 10% Reoplex 400 on Chromosorb WHP (100–120 mesh) at 70 °C and using flame ionization detection.

REFERENCES
1. National Institute for Occupational Safety & Health, 'Manual of Analytical Methods', 2nd ed., 'NIOSH Monitoring Methods', DHEW (NIOSH), Vol. 4, No. 272, US Dept. of Health, Education & Welfare, 1984.
2. R. A. Glaser and W. J. Woodfin, *Am. Ind. Hyg. Assoc. J.*, 1981, **42**, 18.
3. K. Andersson *et al.*, *Chemosphere*, 1983, **12**, 377.

Acrylonitrile

1. CA REGISTRY NO.
107-13-1

2. SYNONYMS
2-Propene nitrile; cyanoethane; cyanoethylene; propene nitrile; vinyl cyanide.

3. MANUFACTURE
3.1 Vapour phase catalytic air oxidation of propylene and ammonia – this is the Sohio Process and is the process in main use today.
3.2 Catalytic oxidation of propylene to acrolein, followed by a catalysed reaction between ammonia and acrolein.
3.3 The ammoxidation of propylene in presence of Te–Ce–Mo oxides at 250 °C, 200 kPa pressure.
3.4 The ammoxidation of propylene in the presence of Mo–V catalyst.
3.5 Liquid phase reaction between acetylene and hydrogen cyanides.
3.6 Decomposition reaction of ethylenecyanohydrin.
3.7 Dehydration of propionitrile in the vapour phase at 600 °C over a chromium oxide catalyst.
3.8 Vapour phase reaction between ammonia and propionaldehyde at 600 °C.
3.9 Uncatalysed reaction between ethylene, propane and butane and hydrogen cyanide at 1000 °C.

4. USES
4.1 Production of acrylic fibre.
4.2 Production of copolymer resins for use in engineering related materials.
4.3 Production of nitrile rubbers and resins in chemical engineering.

5. DETERMINATION IN WORKSHOP ATMOSPHERE
5.1 Recommended Sampling Method

Principle	: Collection and concentration on activated charcoal
Collection tube	: 7 cm × 4 mm i.d.
Absorbant	: 150 mg activated charcoal divided into two parts – 100 mg in the front section separated by a small plug of urethane foam from 50 mg in the rear
Sampling flow	: 10–200 ml min^{-1} by means of a small pump to give a sample of between 4 and 20 l
Extraction	: By means of a 49:1 carbon disulphide–acetone solution

5.2 Recommended Measuring Method

Technique	: GC
Column	: 3 m × 3 mm i.d. stainless steel
Support phase	: Supelcoport (80–100 mesh)
Liquid phase	: 20% SP-1000

Temperature	: Isothermal at 85 °C
Carrier gas	: N_2 or He
Carrier gas flow	: 25 ml min^{-1}
Detector	: FID
Sample size	: 2 µl
Calibration	: solution of distilled acrylonitrile in hexane

5.3 Performance Characteristics

Range studied	: None quoted
Bias	: None identified
Precision	: Not evaluated, but SD of measurement = 0.06
LOD	: 0.001 mg per sample
Interferences	: None known

More detailed information is given in ref. 1.

6. OTHER METHODS

A number of the alternative methods are a modification of the NIOSH method.

6.1 Acrylonitrile was determined in the air of a plastic processing plant by the same charcoal trapping method but the GC analysis was done at 150 °C on a column of 20% TCEP on Chromosorb P.[2]

6.2 Another NIOSH modification was to use methanol to extract the acrylonitrile from the charcoal and to conduct the GC at 140 °C on a column of Poropak Q.[3]

6.3 In a polystyrene manufacturing plant acrylonitrile was trapped using the NIOSH procedure – dimethylformamide was used as an extractant whilst the GC was carried out at 120 °C (on a column of 10% Carbowax on Chromosorb W, the LOD claimed is 1.25 mg m^{-3}).[4]

6.4 Studies of workshop atmospheres at different humidities were carried out using a 600 mg charcoal trap on a 10 ft × 0.125 in stainless steel tube packed with 20% Carbowax + 2% potassium hydroxide on Supelcoport 80–100 mesh at 80 °C with a N–P detector.[5]

6.5 Tenax GC, with thermal desorption, has been used for studies in the home[6] or in stack emissions.[7] Other studies deal with variations in flow rate and desorption temperature.[8]

6.6 A study of 28 compounds, including acrylonitrile, were trapped on a 15 cm × 0.25 in column of Tenax GC from the air at a flow rate of 10–15 ml min^{-1}; after thermal desorption, the acrylonitrile content was determined by GC on a 50 m capillary column coated with SP2100 using a temperature programme of -90 °C to 140 °C; detection was by ECD and the LOD was 0.01 ppb.[9]

6.7 Poropak N has been used as an absorbent with air sampled at 10–15 ml min^{-1} to give a 3 to 8 l sample volume. After thermal desorption, analysis was by GC on a 10 ft × 0.125 in column packed with Poropak N, using a FID detector.[10]

6.8 Another alternative to charcoal is Ambersorb XE-348 (210 mg)[5] and industrial site monitoring used Ti(IV)-exchanged Zeolite 3A resins; this resin concentrated the acrylonitrile by conversion to polyacrylonitrile.[11]

6.9 Direct injection into a 10 ft × 0.125 in stainless steel column containing 20% SP-1200 with 1% Carbowax 1500 as support phase;[12] a N–P detector was used.

6.10 The toxic volatiles, from burning plastic, were sampled directly and separated on a 2 m × 2 mm column packed with Poropak Q; operating temperature was 163 °C and a FID was used.[13]

6.11 In a similar experiment to **6.10** the toxic volatiles from N containing polymers were examined by GC on a 2 m × 3.2 mm stainless steel column packed with Poropak Q. Temperature programming from 30 to 250 °C at 8 °C min^{-1} was used and detection was by flame ionization detector connected to a chemiluminescent device which measured the oxidation of NO with ozone.[14] A chemiluminescent detector has also been used directly for determining acrylonitrile (and others) in ambient air but the technique lacks sensitivity.[15]

6.12 In general studies of organic volatiles infrared measurements have been made, either with a CO_2 laser[16] or after separation by HRGC.[17]

6.13 One of the earliest attempts to measure the acrylonitrile content of air was an assessment of the 'knock-down' of *Tribolium castaneum* when acrylonitrile has been used as a grain fumigant.[18]

REFERENCES

1. National Institute for Occupational Safety & Health, 'Manual of Analytical Methods', 3rd ed., 'NIOSH Monitoring Methods', DHEW (NIOSH) Publications No. 1604, US Dept. of Health, Education & Welfare, 1984.
2. D. R. Marrs *et al.*, Plast. Packag./Acrylonitrile, Natl. Tech. Conf. Soc. Plat. Eng., 1978, p. 131.
3. Z. Renton, *VIA, Varian Instrum. Appl.*, 1979, **13**(2), 4.
4. H. Tyras and J. Stufka-Olczyk, *Chem. Anal. (Warsaw)*, 1984, **29**, 281.
5. R. G. Melcher *et al.*, *Am. Ind. Hyg. Assoc. J.*, 1986, **47**, 152.
6. F. H. Jarke and S. M. Gordon, Proc. Annu. Meet. Air Pollut. Control Assoc. 4, 1981, 81-57.2.
7. J. S. Parsons and S. Mitzner, *Environ. Sci. Technol.*, 1975, **9**, 1053.
8. R. H. Brown and C. J. Purnell, *J. Chromatog.*, 1979, **178**, 79.
9. B. B. Kebbekus and J. W. Bozzelli, Proc. Annu. Meet. Air Pollut. Control Assoc. 4, 1982, 82-65.2.
10. R. Dan and R. H. Moore, *Am. Ind. Hyg. Assoc. J.*, 1979, **40**, 904.
11. M. L. Markowski, *Diss. Abstr. Int. B*, 1985, **46**, 823.
12. S. W. Cooper *et al.*, *J. Chromatog. Sci.*, 1986, **24**, 204.
13. D. S. Duvall and W. A. Rubey, West. States Sect., Combust. Inst. (Pap.) WSCI, 1973, 73.
14. J. M. Murrell, *Fire Matr.*, 1986, **10**(2), 57.
15. J. C. Hilborn *et al.*, Int. Conf. Environ. Sensing Assess. (Proc.) 2, 1975, 3pp.
16. D. M. Sweger and J. C. Travis, *Appl. Spectrosc.*, 1979, **33**, 46.
17. G. T. Reedy *et al.*, *Anal. Chem.*, 1985, **57**, 1602.
18. M. Muthu *et al.*, *Int. Pest Contr.*, 1971, **13**(4), 11.

o-Tolidine

1. CA REGISTRY NO.
119-93-7

2. SYNONYMS
3,3'-dimethylbiphenyl 4,4'-diamine;
3,3'-dimethylbenzidine.

3. MANUFACTURE
Reduction of nitrobenzene, in alkaline media, using Zn or methanol to give hydrazobenzene – after separation this is re-arranged to give benzidine, and controlled methylation to give *o*-tolidine.

4. USES
Used mainly in the direct production of azo dyes.

5. DETERMINATION IN WORKSHOP ATMOSPHERE
5.1 Recommended Sampling Method
Technique	: Trapping on membrane filter
Adsorber	: 37 mm diam. 5 μm PTFE membrane filters
Sample flow	: 1 to 3 l min^{-1}
Volume	: 150 l (at 0.1 mg m^{-3}) to 500 l
Extraction	: 1. Place filter in 50 ml beaker, face up
	2. Add 1 ml H$_2$O and swirl
	3. Add 1 ml H$_2$O and shake
	4. Turn filter over and place beaker in ultrasonic bath for 15 min
	5. Transfer 1 ml aliquot to 4 ml vial
	6. Add 1 ml of a solution containing 200 mg Na$_2$S$_2$O$_4$ in 10 ml phosphate buffer [1.179 g KH$_2$PO$_4$ + 4.3 g Na$_2$HPO$_4$ in 1 l water]
	7. Cap, and stand for 1 h with occasional agitation.

5.2 Recommended Measuring Method
Technique	: HPLC
Column	: 10 cm × 8 mm i.d.
Active phase	: 10 μm Waters Radial Pak C$_{18}$
Mobile phase	: 60/40 methanol/phosphate buffer [buffer is 3.39 g KH$_2$PO$_4$ + 4.3 g Na$_2$HPO$_4$ in 1 l of water]
Temperature	: Ambient
Detector	: U.v. at 280 nm
Flow	: Not given
Aliquot size	: 10 μl

5.3 Performance Characteristics

Range studied	: Not studied but the method works from 15 to 250 μg per sample
Bias	: None identified
Overall precision	: Not evaluated but SD of measurement = 0.04 to 0.08
Interferences	: None identified

More detailed information is given in ref. 1. This method, ref. 1, is based on work described in ref. 2.

6. OTHER METHODS

6.1 There has been only a small interest in the determination of o-tolidine in air. An early method involved trapping several different amines by bubbling the vapour through 0.1N HCl and taking an aliquot of the solution for colorimetric determination.[3] The aliquot was made alkaline and extracted with diethyl ether and the extract was dissolved in 10% HCl and subjected to paper chromatography, and the resulting band containing the amines was eluted with N HCl. The amines were reacted with N-(1-naphthyl) ethylenediamine and the resulting colour was measured at 575 nm. The limit of detection was 5 μg of amine.

6.2 An alternative HPLC method to the NIOSH one collects o-tolidine from air at a flow-rate of 1.5–2 m^3 min^{-1} (to give a sample of 2000 m^3) on a glass fibre filter.[4] The collected particulate was extracted with 0.1M sodium phosphate buffer (pH 7.0) plus chloroform, and after shaking and separation, the chloroform layer was concentrated. After a clean-up of the extract and the addition of methanol and 0.1M sodium acetate buffer (pH 4.7) an aliquot was taken for analysis on a 25 cm × 4.6 mm column packed with LiChrosorb RP-2 (5 μm). The mobile phase was acetonitrile–0.2M sodium acetate buffer (pH 4.7) (1:1) and was used at 1 ml min^{-1}; detection was by electrochemical detector equipped with a glassy carbon electrode. The limit of detection was 1 ng m^{-3} but the recovery was only 30–45% at a spiking level of 25 ng m^{-3} of air. The method can also be used for the determination of o-dianisidine and 3,3′-dichlorobenzidine.

6.3 In a GC method, benzidine and its derivatives, i.e., o-tolidine, o-dianisidine and 3,3′-dichlorobenzidine, were converted to their pentafluoropropionamides by reaction with N-pentafluoropropionylimidazole in toluene.[5] They were separated on a 6 ft × 2 mm glass column packed with a 9:9:2 mixture of 3% OV-17 on Chromosorb W AW DMCS (100–120 mesh), 3% OV-210 on Chromosorb W AW DMCS (80–100 mesh) and 3% OV-225 on Chromosorb W HP (100–120 mesh); detection was by electron capture. Although the method was applied to aqueous samples, the limit of detection being less than 1 μg l^{-1}, the GC conditions could be suitable to apply to aqueous extracts of particulate matter collected from workshop air.

6.4 Pentafluoropropionic anhydride was used to prepare derivatives of aromatic amines, including o-tolidine and o-dianisidine[6] to give the same products for GC analysis as in the previous reference. A 30 m capillary column coated with OV-73 was used to separate the derivatives by temperature programming up to 290 °C and a thermionic detector was used in the N mode. If there is a need to separate the benzidine derivatives before their determination, this method is the one of choice.

6.5 Having trapped amines, including o-tolidine, and isocyanates from air by bubbling through dilute acid, and after making the solution alkaline and extracting the amines with toluene, a similar method involving the GC of the pentafluoropropionamides was used.[7] In this method the detection was by ^{63}Ni electron capture, the limit of detection being 1 pg.

REFERENCES

1. National Institute for Occupational Safety & Health, 'Manual of Analytical Methods', 3rd ed., 'NIOSH Monitoring Methods', DHEW (NIOSH) Publication No. 5013, US Dept. of Health, Education & Welfare, 1984.
2. Development of an Analytical Method for Benzidine-based dyes; E. R. Kennedy and M. J. Seymour, 'Chemical Hazards in the Workplace: Measurement and Control', *Am. Chem. Soc. Symp. Ser.*, 1981, No. 149.
3. G. Ghetti *et al.*, *Lav. Um.*, 1968, **20**, 389.
4. R. M. Riggin *et al.*, *J. Chromatog. Sci.*, 1983, **21**, 321.
5. F. K. Kawahara *et al.*, *Anal. Chim. Acta*, 1982, **138**, 207.
6. G. Skarping *et al.*, *J. Chromatog.*, 1983, **270**, 207.
7. G. Skarping *et al.*, *J. Chromatog.*, 1983, **267**, 315.

o-Dianisidine and Salts

1. CA REGISTRY NO.
119-90-4

2. SYNONYMS
3,3'-Dimethoxybenzidine.

3. MANUFACTURE
Catalytic or electrolytic alkoylation of benzidine.

4. USES
Primarily used in the direct preparation of azo dyes.

5. DETERMINATION IN WORKSHOP ATMOSPHERE
5.1 Recommended Sampling Method

Technique	: Trapping and concentration on membrane filter
Adsorber	: 5 μm PTFE membrane 37 mm diam. and personal cassette holder
Sample flow	: 1 to 3 l min^{-1}
Sample volume	: 150 l (at 0.1 mg m^{-3}) to 500 l
Extraction	: 1. Place filter face-up in 50 ml beaker
	2. Add 1 ml H$_2$O and swirl to wet the deposit
	3. Add 1 ml H$_2$O and shake
	4. Reverse filter (sample side down) and place in an ultrasonic bath for 15 min
	5. Transfer 1 ml aliquot to 4 ml vial and add 1.0 ml of reducing solution (200 mg Na$_2$S$_2$O$_4$ + 10 ml of a solution containing 1.179 g KH$_2$PO$_4$ in 1 l H$_2$O) freshly prepared
	6. Cap vial and stand for 1 h with occasional agitation

5.2 Recommended Measuring Method

Technique	: HPLC
Column	: 10 cm × 8 mm i.d.
Support phase	: 10 μm particles of Waters Radial Pak C$_{18}$ or equivalent
Liquid phase	: 60% methanol: 40% phosphate buffer (phosphate buffer is 3.39 g KH$_2$PO$_4$ + 4.3 g Na$_2$HPO$_4$ diluted to 1 l with H$_2$O)
Temperature	: Ambient
Detector	: U.v. at 280 nm
Carrier gas	: N/A
Carrier gas flow	: Not given
Sample size	: 10 μl

5.3 Performance Characteristics

Range studied	: 0.06 to 8 mg m^{-3} on a 250 l air sample
Bias	: None identified
Overall precision	: Not evaluated, but precision of measurement SD = 0.04 to 0.08
LOD	: 3 µg sample
Interferences	: None detected

More detailed information is given in ref. 1 – this method is based on work of an earlier publication.[2]

6. OTHER METHODS

6.1 There has been only a small interest in the determination of o-dianisidine in air. An early method involved trapping several different amines by bubbling the vapour through 0.1N HCl and taking an aliquot of the solution for colorimetric determination.[3] The aliquot was made alkaline and extracted with diethyl ether and the extract was dissolved in 10% HCl and subjected to paper chromatography and the resulting band containing the amines was eluted with N HCl. The amines were reacted with N-(1-naphthyl)ethylenediamine and the resulting colour was measured at 575 nm. The limit of detection was 5 µg of amine.

6.2 An alternative HPLC method to the NIOSH one collects o-dianisidine from air at a flow-rate of 1.5–2 m^3 min^{-1} (to give a sample of 2000 m^3) on a glass fibre filter.[4] The collected particulate was extracted with 0.1M sodium phosphate buffer (pH 7.0) plus chloroform and, after shaking and separation, the chloroform layer was concentrated. After a clean-up of the extract and the addition of methanol and 0.1M sodium acetate buffer (pH 4.7) an aliquot was taken for analysis on a 25 cm × 4.6 mm column packed with LiChrosorb RP-2 (5 µm). The mobile phase was acetonitrile–0.2M sodium acetate buffer (pH 4.7) (1:1) and was used at 1 ml min^{-1}; detection was by electrochemical detector equipped with a glassy carbon electrode. The limit of detection was 1 ng m^{-3} but the recovery was only 30–45% at a spiking level of 25 ng m^{-3} of air. The method can also be used for the determination of o-tolidine and 3,3′-dichlorobenzidine.

6.3 Three GC methods have been used. Benzidine and its derivatives, i.e., o-dianisidine, o-tolidine and 3,3′-dichlorobenzidine, were converted to their pentafluoropropionamides by reaction with N-pentafluoropropionylimidazole in toluene.[5] They were separated on a 6 ft × 2 mm glass column packed with a 9:9:2 mixture of 3% OV-17 on Chromosorb W AW DMCS (100–120 mesh), 3% OV-210 on Chromosorb W AW DMCS (80–100 mesh), 3% OV-225 on Chromosorb W HP (100–120 mesh); detection was by electron capture. Although the method was applied to aqueous samples, the limit of detection being less than 1 µg l^{-1}, the GC conditions could be suitable to apply to aqueous extracts of particulate matter collected from workshop air.

6.4 Pentafluoropropionic anhydride was used to prepare derivatives of aromatic amines, including o-dianisidine and o-tolidine,[6] to give the same products for GC analysis as in the previous reference. A 30 m capillary column coated with OV-73 was used to separate the derivatives by temperature programming up to 290 °C and a thermionic detector was used in the N mode. If there is a need to separate the benzidine derivatives before their determination, this method is the one of choice.

6.5 In a field test method for the determination of primary aromatic amines in air, 1 l samples at 250 ml min^{-1} were taken and o-anisidine, o-tolidine or 3,3'-dichlorobenzidine were trapped in a bubbler containing 0.1M HCl.[7] To the solution was added heptafluorobutyryl chloride in cyclohexane and also pyridine and the mixture was shaken and the cyclohexane layer examined by GC. This was performed on a 1.5 m × 6 mm column packed with 2% SE-52 on Chromosorb W (80–100 mesh) operated at 210 °C for o-dianisidine; detection was by electron capture detector.

REFERENCES

1. National Institute for Occupational Safety & Health, 'Manual of Analytical Methods', 3rd ed., 'NIOSH Monitoring Methods', DHEW (NIOSH) Method No. 5013, US Dept. of Health, Education & Welfare, 1984.
2. E. R. Kennedy and M. J. Seymour 'Chemical Hazards in the Workplace: Measurement and Control', *Am. Chem. Soc. Symp. Ser.*, 1981, No. 149, 21.
3. G. Ghetti *et al.*, *Lav. Um.*, 1968, **20**, 389.
4. R. M. Riggin *et al.*, *J. Chromatog. Sci.*, 1983, **21**, 321.
5. F. K. Kawahara *et al.*, *Anal. Chim. Acta*, 1982, **138**, 207.
6. G. Skarping *et al.*, *J. Chromatog.*, 1983, **270**, 207.
7. D. W. Meddle and A. E. Smith, *Analyst*, 1981, **106**, 1082.

3,3'-Dichlorobenzidine

1. CA REGISTRY NO.
91-94-1

2. SYNONYMS
3,3'-dichlorobiphenyl-4,4'-diamine.

3. MANUFACTURE
By the halogenation of benzidine.

4. USES
Mainly in the production of HANS yellow and diarylide yellow dyes.

5. DETERMINATION IN WORKSHOP ATMOSPHERE
5.1 Recommended Sampling Method
Technique	: Filter and solid sorbent tube
Filter	: 13 mm glass fibre
Adsorbent	: 50 ng (30–60 mesh) silica gel in a 3 cm × 4 mm glass tube
Sample flow	: 0.2 l min^{-1}
Sample volume	: 20 l at 10 µg m^{-3} to 100 l
Extraction	: 1. Place filter and silica gel in separate test tubes
	2. Add 0.5 ml of 0.17% triethylamine in methanol to each
	3. Stand for 1 h with intermittent shaking
	4. Centrifuge each sample for 10 min

5.2 Recommended Measuring Method
Technique	: HPLC
Column	: 30 cm × 4 mm
Stationary phase	: 10 µm Bondapak C_{18}
Mobile phase	: 70% acetonitrile/30% water
Flow rate	: 1.5 ml min^{-1}
Temperature	: Ambient
Detector	: U.v. at 254 nm
Calibration	: Solution of 3,3'-dichlorobenzidine

5.3 Performance Characteristics
Range studied	: 20 to 130 µg m^{-3}
Bias	: Not determined
Overall precision	: SD = 0.07
LOD	: 0.05 µg per sample
Interferences	: 4,4'-methylenebis (2-chloroaniline)

More detailed information is given in ref. 1.

6. OTHER METHODS

6.1 This recommended method is based on a previously published method.[2,3] The glass fibre filter traps the dust and fume containing 3,3'-dichlorobenzidine whereas the silica traps the vapour.

6.2 This trapping method is an improvement on the use of impingers and bubblers.[4,5] Among compounds which interfere in the method is 4,4'-methylenebis (2-chloroaniline).

6.3 An early method used trapping of 3,3'-dichlorobenzidine and other amines in 0.1N HCl and extracting the solution with diethyl ether after making alkaline.[6] The extract was evaporated and the residue dissolved in 10% HCl and an aliquot transferred to chromatographic paper and the amine bands were developed with isobutanol–acetic acid–water. N-(1-naphthyl)-ethylenediamine was used to identify the bands, and also as reactant with the amines extracted from the paper with N HCl, measuring the absorbance at 575 nm. A limit of detection of 5 μg amine was claimed.

6.4 An alternative HPLC method to the NIOSH one collects 3,3'-dichlorobenzidine from air at a flow-rate of $1.5–2 \, m^3 \, min^{-1}$ (to give a sample of $2000 \, m^3$) on a glass fibre filter.[7] The collected particulate was extracted with 0.1M aqueous sodium phosphate buffer (pH 7.0) plus chloroform and after shaking and separation, the chloroform was concentrated. After a clean-up of the extract and the addition of methanol and 0.1M sodium acetate buffer (pH 4.7) an aliquot was taken for analysis on a 25 cm × 4.6 mm column packed with LiChromosorb RP-2 (5 μm). The mobile phase was acetonitrile–0.2M sodium acetate buffer (pH 4.7) (1:1) and was used at 1 ml min^{-1}; detection was by electrochemical detector equipped with a glassy carbon electrode. The limit of detection was $0.1–1 \, ng \, m^{-3}$. The method can also be used for the determination of o-dianisidine and o-tolidine.

6.5 In a GC method, benzidine and its derivatives, *i.e.*, 3,3'-dichlorobenzidine, o-dianisidine and o-tolidine, were converted to their pentafluoropropionamides by reaction with N-pentapropionylimidazole in toluene.[8] They were separated on a 6 ft × 2 mm glass column packed with a 9:9:2 mixture of 3% OV-17 on Chromosorb W AW DMCS (100–120 mesh), 3% OV-210 on Chromosorb W AW DMCS (80–100 mesh) and 3% OV-225 on Chromosorb W HP (100–120 mesh); detection was by electron capture. Although the method was applied to aqueous samples, the limit of detection being less than $1 \, \mu g \, l^{-1}$, the GC conditions could be suitable to apply to aqueous extracts of particulate matter collected from workshop air.

6.6 In a field test for the determination of primary aromatic amines in air, 1 l samples, at 250 ml min^{-1}, were taken and 3,3'-dichlorobenzidine, o-dianisidine or o-tolidine were trapped in a bubbler containing 0.1M HCl.[9] To the solution was added heptafluorobutyryl chloride in cyclohexane and also pyridine and the mixture was shaken, the cyclohexane layer being examined by GC. This was performed on a 1.5 m × 6 mm column packed with 2% SE-52 on Chromosorb W (80–100 mesh) operated at 230 °C for 3,3'-dichlorobenzidine; detection was by electron capture detector.

REFERENCES
1. National Institute for Occupational Safety & Health, 'Manual of Analytical Methods', 3rd ed., 'NIOSH Monitoring Methods', DHEW (NIOSH) Method No. 5509, US Dept. of Health, Education & Welfare, 1984.
2. R. Morales and R. E. Hermes, *Am. Ind. Hyg. Assoc. J.*, 1979, **40**, 970.
3. R. Morales *et al.*, *IARC Sci. Publ.*, 1981, **40**, 119.
4. J. M. Glassman and J. W. Meigs, *A.M.A. Arch. Ind. Hyg. Occup. Med.*, 1951, **4**, 519.
5. L. T. Butt and N. Strafford, *J. Appl. Chem.*, 1956, **6**, 525.
6. G. Ghetti *et al.*, *Lav. Um.*, 1968, **20**, 389.
7. R. M. Riggin *et al.*, *J. Chromatog. Sci.*, 1983, **21**, 321.
8. F. K. Kawahara *et al.*, *Anal. Chim. Acta*, 1982, **138**, 207.
9. D. W. Meddle and A. E. Smith, *Analyst*, 1981, **106**, 1082.

4,4'-Methylene Bis(2-Chloroaniline)

1. CA REGISTRY NO.
101-14-4

2. SYNONYMS
MBOCA; MOCA;
di(4-amino-3-chlorophenyl) methane;
DACPM;
4,4'-diamino-3,3'-dichlorodiphenylmethane; methylenebis (o-chloraniline);
p,p'-methylenebis (o-chloraniline).

3. MANUFACTURE
Reaction of formaldehyde with o-chloraniline.

4. USES
4.1 Curing agent for isocyanate containing polymers.
4.2 Curing agent for liquid-castable polyurethane elastomers.

5. DETERMINATION IN WORKSHOP ATMOSPHERE
5.1 Recommended Sampling Method
Technique	: 2-Stage sampler glass fibre filter and silica gel
Adsorber	: 13 mm diam. glass fibre filter (Gelman Type A or equivalent) connected to a sorbent tube (3 cm × 4 mm i.d.) containing 50 mg 30–60 mesh silica gel
Flow	: 0.2 to 1.0 l min^{-1} by means of personal pump
Sample volume	: 15 min to 8 h at flow rates given above
Extraction	: 1. Place glass fibre filter and silica gel in same test tube
	2. Add 0.5 ml methanol, seal, shake and stand for 1 h with intermittent agitation; centrifuge for 10 min

5.2 Recommended Measuring Method
Technique	: HPLC
Column/support	: μBondapak C_{18} 0.63 cm i.d. × 30 cm
Liquid phase	: 8:2 (v/v) acetonitrile:water
Temperature	: Column 22 °C; injector 22 °C
Column pressure	: 1400 p.s.i.
Detector	: U.v. at 245 nm
Carrier gas	: N/A
Sample flow	: 1.3 ml min^{-1}
Sample size	: 10 μl

4,4'-Methylene Bis(2-Chloroaniline) 53

5.3 Performance Characteristics

Range studied : 3 to 200 µg m^{-3} in a 50 l sample
Bias : None quoted
Precision : $CV_T = 0.1$ at 49 µg m^{-3}
LOD : 0.15 µg (= 3 µg m^{-3} in 50 l sample)
Interferences : α- and β-naphthlylamine, *o*-tolidine,
N-methylaniline,
3,3'-dichlorobenzidine,
2-chloro-4-methylaniline,
p-tolidine, and 3-chloromethylaniline: high
relative humidity causes the trapping efficiency of
silica gel to be reduced

More detailed information is given in ref. 1.

6. OTHER METHODS

6.1 Similar compounds, *e.g.*, *o*-tolidine and 3,3'-dichlorobenzidine interfere in the method, which is based on a previously published one[2] and subsequently slightly modified.[3] If it is necessary to separate these compounds before analysis, a gas chromatographic (GC) method is recommended (see later).

6.2 In an alternative HPLC method, air was sampled at 1 l min^{-1} for 40 min through a GFA-type glass fibre filter and a tube containing 30 mg of Tenax GC (35–60 mesh).[4,5] Methanol was used to extract the MOCA from both trapping media and the extract was analysed on a 15 cm × 4.5 mm column packed with Hypersil ODS (5 µm) using methanol–sodium phosphate buffer (pH 7.5) (3:2) as mobile phase at a flow-rate of 1.5 ml min^{-1}. Electrochemical detection with a glassy carbon working electrode was used as a more selective alternative to u.v. detection, the limit of detection being 100 ng m^{-3} air.

6.3 Ethanol made alkaline with potassium hydroxide was used to trap amines, including MOCA, and aromatic isocyanates from workshop atmospheres.[6] After a reaction to separate isocyanates an aliquot of the resulting 50% aqueous ethanolic solution was used for HPLC analysis on a Radial-PAK C_{18} column. Tetrahydrofuran–acetonitrile–sodium acetate buffer (pH 5.5–7.0) (3:3:4) was used as mobile phase and detection was by u.v. at 245 nm. The detection limit was 1–5 µg m^{-3} for a 10 l sample.

6.4 GC has also been used to separate and determine MOCA in workshop atmospheres. Air was trapped at 1 l min^{-1} for up to 8 h through a tube containing 50 mg of Gas Chrom S and a back-up tube containing 10 mg of the same material (separated by a quartz plug).[7] Acetone was used to extract the MOCA from the combined silica samples and GC was carried out on the extract using a 1 ft × 0.125 in (o.d.) stainless steel column packed with 10% Dexsil 300 GC on ABS Anakrom (80–90 mesh). The column temperature was 200 °C and detection was by flame ionization, the limit of detection being 2 µg for 480 l of air.

6.5 A similar GC method was published in the same period.[8]

6.6 MOCA was determined simultaneously with 2,4- and 2,6-di-isocyanatotoluene after passing contaminated air at 1 l min^{-1} through a midget impinger containing a mixture of acetic acid–water–hydrochloric acid (5:4:1).[9] The solution was made alkaline and extracted with chloroform prior to formation of the trifluoroacetamide derivatives. The toluene solution of these was subjected to GC on a 1 m × 2 mm column packed with 3% OV-101 on Chromosorb WHP (60–80 mesh). The column was temperature programmed from 120 to 240 °C at 20 °C min^{-1}. Detection was by ^{63}Ni electron capture and the limit of detection was 0.2 ng.

6.7 Amines, including MOCA, and isocyanates were trapped from air in dilute acid and the resulting solution was made alkaline and the amines were extracted with toluene.[10] Pentafluoropropionic anhydride was added and the resulting derivatives were separated by GC on a 25 m × 0.32 mm glass column coated with OV-73. The column was temperature programmed up to 300 °C at 10 °C min^{-1} and detection was by ^{63}Ni electron capture, the limit of detection being 1 pg.

6.8 A similar method was published in the same period but using thermionic detection in the N mode.[11]

6.9 Pentafluoropropionamides were used in a similar method after trapping primary and secondary amines and isocyanates from air into 0.2% ethanolic potassium hydroxide solution.[12] The icocyanates were converted to urethanes by the potassium hydroxide leaving the amines to be converted to their pentafluoropropionamides after extraction by toluene and reaction with pentafluoropropionic anhydride. GC was carried out on the derivatives on a 10 m or 15 m × 0.32 mm glass column coated with PS-225 cross linked with azo-t-butane. Thermionic detection in the N mode was used, the limit of detection for the MOCA derivative being in the range 40–80 fmol. If it is necessary to separate MOCA from o-tolidine and 3,3′-dichlorobenzidine, this GC method is the one that is recommended.

REFERENCES

1. National Institute for Occupational Safety & Health, 'Manual of Analytical Methods', 2nd ed., 'NIOSH Monitoring Methods', DHEW (NIOSH) Method No. 236, US Dept. of Health, Education & Welfare, 1977.
2. R. Morales et al., 'Development of sampling and analytical methods for carcinogens'. July 1–Dec 31, 1987, LA-6387-PR Los Alamos Scientific Lab. 1975.
3. S. M. Rappaport and R. Morales, *Anal. Chem.*, 1979, **51**, 19.
4. C. J. Purnell and C. J. Warwick, *Analyst*, 1980, **105**, 861.
5. C. J. Purnell and C. J. Warwick, *IARC Sci. Publ.*, 1981, **40**, 133.
6. E. H. Nieminen et al., *J. Liq. Chromatog.*, 1983, **6**, 453.
7. S. K. Yasuda, *J. Chromatog.*, 1975, **104**, 283.
8. E. Sawicki, *Health Lab. Sci.*, 1975, **12**, 415.
9. G. F. Ebell et al., *Ann. Occup. Hyg.*, 1980, **32**, 185.
10. G. Skarping et al., *J. Chromatog.*, 1983, **267**, 315.
11. G. Skarping et al., *J. Chromatog.*, 1983, **270**, 207.
12. G. Skarping et al., *J. Chromatog.*, 1985, **346**, 191.

2-Nitronaphthalene

1. CA REGISTRY NO.
581-89-5

2. SYNONYMS
Naphthalene, 2-nitro;
β-nitronaphthalene.

3. MANUFACTURE
Via the Bucherer reaction starting with 2-naphthalenol.

4. USES
No known uses, appears to be a by-product in manufacture.

5. DETERMINATION IN WORKSHOP ATMOSPHERE
5.1 Recommended Sampling Method

Technique	: Collection and concentration on Teflon coated glass fibre filter
Filter	: 47 mm diam. Teflon coated glass fibre filter
Sample flow	: Not $> 1 \, l \, min^{-1}$
Sample volume	: Sufficient to collect a minimum of 10 mg of particulates
Extraction	: 1. The filter, plus particulates, is extracted for 16 h in darkness with 250 ml dichloromethane
	2. The extract is evaporated to dryness under nitrogen
	3. Residue is dissolved in 1:1 methanol–dichloromethane
	4. Solution passed through a catalytic 7.6 cm × 4.6 mm column of 5 μm alumina coated with Pt–Rh catalyst – the mobile phase is aqueous methanol (70%) and flow rate is $1 \, ml \, min^{-1}$. Resulting amine is passed on to measuring column (see 5.2 below)

5.2 Recommended Measuring Method

Technique	: HPLC
Column	: 15.0 cm × 4.6 mm
Support phase	: Zorbax ODS
Mobile phase	: 4:1 methanol–water
Temperature	: Ambient
Detection	: Fluorescence spectrophotometer – excitation at 234 nm and detection at 403 nm
Carrier gas	: N/A
Flow	: $1 \, ml \, min^{-1}$

5.3 Performance Characteristics

Range studied : Not quoted
Bias : None quoted
Precision : RSD = 2.7%
LOD : 14 pg
Interference : None quoted

More detailed information is given in ref. 1.

6. OTHER METHODS

6.1 A similar sampling and extraction method was used for 2-nitronaphthalene in diesel exhaust fumes.[2] A 4.7 mm diameter Teflon coated fibre filter was used and the collected particulate was extracted with dichloromethane. Various clean-up procedures were used including passage of the extract through a silica cartridge and HPLC on a 30 cm × 7.8 mm column packed with silica (10 μm). The mobile phase was hexane–dichloromethane (99:1) for 10 min changing, on a linear programme, to dichloromethane, holding for 20 min and then returning to the original solvent composition. The flow-rate was 1.5 ml min^{-1} and detection was at 254 nm, the nitro polyaromatic hydrocarbons, including 2-nitronaphthalene being eluted between 23 and 75 min. The eluted extract was analysed by gas chromatography (GC) on a 30 m capillary column coated with 5PB-5 silicone (0.25 μm). After temperature-programming, 2-nitronaphthalene was detected by electron capture, the limit of detection being 2 ppb.

6.2 Capillary GC has also been used for the analysis of nitrated polynuclear aromatic hydrocarbons.[3] Stationary phases of 5% phenyl methyl silicone and polyethylene glycol were used and flame ionization detection.

6.3 2-Nitronaphthalene seldom occurs on its own in air and the accompanying compounds cause clean-up problems when using GC analysis, which have to be overcome by chromatographic procedures[2,4] or by using mass spectrometric detection.[5-7]

6.4 In an examination of the excitation and phosphorescence spectra of 22 compounds tentatively identified in diesel exhaust fumes, 2-nitronaphthalene was included[8] although 1-nitropyrene was the pollutant of prime consideration.

6.5 The same compound prompted an investigation into surface-enhanced Raman spectroscopy as an analytical technique with 2-nitronaphthalene included in the list of compounds covered.[9]

REFERENCES

1. S. B. Tejada *et al.*, *Anal. Chem.*, 1986, **58**, 1827.
2. W. M. Draper, *Chemosphere*, 1986, **15**, 437.
3. H. Matsushita *et al.*, *Taiki Osen Gakkaishi*, 1983, **18**, 241.
4. P. Ciccioli and A. Liberti, *Riv. Combust.*, 1985, **39**(4–5), 111.
5. A. Liberti *et al.*, *J. High Resolut. Chromatog., Chromatog. Commun.*, 1984, **7**, 389.

6. J. Tuominen *et al.*, *J. High Resolut. Chromatog., Chromatog. Commun.*, 1986, **9**, 469.
7. U. Sellstroem *et al.*, *Chemosphere*, 1987, **16**, 945.
8. O. S. Wolfbeis *et al.*, *Anal. Chim. Acta*, 1983, **147**, 405.
9. T. Vo-Dinh and P. D. Endow, Proc. APCA Annu. Meet. 78th, 6, 1985, 85-81-3.

5-Nitroacenaphthene

1. CA REGISTRY NO.
602-87-9

2. SYNONYMS
Acenaphthylene, 1,2-dihydro-5-nitro;
acenapthene, 5-nitro.

3. MANUFACTURE
3.1 Synthesized by the nitration of acenaphthene in acetic anhydride solution using Cu, or $Zn(NO_3)_2$, at 30 °C as catalyst.
3.2 Nitration of acenaphthene with nitric acid in sulphuric acid solution.

4. USES
Used as a chemical intermediate in the production of naphthalimide dyes.

5. DETERMINATION IN WORKSHOP ATMOSPHERE
5.1 Recommended Sampling Method
Technique	: Trapping and concentration on quartz fibre filter
Adsorber	: Quartz fibre filter with high volume sampler
Sample flow	: Not quoted
Sample volume	: 1700 m^3 over 24 h
Extraction	: (Details given in ref. 2)

5.2 Recommended Measuring Method
Technique	: GC
Column	: 12.5 m × 0.2 mm fused SiO_2 capillary
Support phase	: Capillary column wall
Active phase	: 5% phenyl methyl silicone chemically bonded to column walls
Temperature	: Column 70 to 300 °C at 6 °C min^{-1}; injector 250 °C; detector 300 °C
Detector	: NPD with H_2 at 3 ml min^{-1}; He at 30 ml min^{-1}; Air at 100 ml min^{-1}
Carrier gas	: He
Carrier gas flow	: 1.0 ml min^{-1}
Sample size	: 0.5 ml (see Note)

Note: Not actually quoted, but arrived at by calculation.

5.3 Performance Characteristics
Range studied	: Note quoted
Bias	: None quoted
Overall precision	: CV = 0.09%
LOD	: 0.05 mg
Interferences	: None quoted

More detailed information is given in ref. 1.

6. OTHER METHODS

The two methods are almost identical and in the event of there being no other suitable method must form the basis of the method that is recommended. For full details of the methods see refs. 1 and 2. Assuming a small sampling only is necessary, the sampling method uses a small portable apparatus with no liquids and is therefore easy to handle and maintain. The analytical method is relatively specific with the additional advantage of flexibility of operating conditions.

REFERENCES
1. H. Matsushita et al., Taiki Osen Gakkaishi, 1983, **18**, 241.
2. H. Matsushita and Y. Iida, J. High Resolut. Chromatog., Chromatog. Commun., 1986, **9**, 708.

Diethyl Sulphate

1. CA REGISTRY NO.
64-67-5

2. SYNONYMS
DES; sulphuric acid diethyl ester; ethyl sulphate.

3. MANUFACTURE
Principal method in use is the absorption of ethylene in sulphuric acid with Sb, Sn or Bi as catalyst.

4. USES
4.1 As a feedstock in the manufacture of ethyl chloride.
4.2 As an intermediate in the manufacture of ethyl alcohol.

5. DETERMINATION IN WORKSHOP ATMOSPHERE
5.1 Recommended Sampling Method

Technique	: Trapping on solid absorbent
Adsorbent	: 20–40 mesh silica gel – 520 mg in front section of tube separated by a polyurethane plug from 260 mg in the rear section
Sample flow	: 200 ml min^{-1} for 100 min by means of small pump
Sample volume	: 20 l
Extraction	: The 780 mg silica gel is stood for 30 min in 10 ml acetone

5.2 Recommended Measuring Method

Technique	: GC
Column	: 10 ft × 0.125 in stainless steel
Support phase	: 80–100 mesh Supelcoport
Liquid phase	: 10% OV-101
Temperature	: 110 °C isothermal
Detector	: Flame photometer at 394 nm
Carrier gas	: N_2
Carrier gas flow	: 80 ml min^{-1}
Sample size	: 5 µl

5.3 Performance Characteristics

Range studied	: 5 to 50 µg ml^{-1}
Bias	: None noted
Overall precision	: ± 38 µg ml^{-1} at 95% confidence limit
LOD	: 0.1 ppm on 20 l sample
Interferences	: None mentioned

More detailed information is given in ref. 1.

6. OTHER METHODS
There is a dearth of information on the determination of diethyl sulphate in air,

only one other method being found.[2] This also uses silica gel to trap diethyl sulphate and other alkly sulphates before extraction of the silica gel with diethyl ether. The extract is analysed by GC using electron capture detection, the limit of detection being 11 µg m^{-3} air.

REFERENCES
1. J. C. Gilland and A. P. Bright, *Am. Ind. Hyg. Assoc. J.*, 1980, **41**, 459.
2. H. Blome and M. Hennig, Ger. Offen. DE3, 400,134.

Dimethyl Sulphate

1. CA REGISTRY NO.
77-78-1

2. SYNONYMS
DMS.

3. MANUFACTURE
The exothermic reaction between gaseous dimethyl ether and liquid sulphur trioxide at approx. 46 °C; the reaction product is purified by vacuum distillation.

4. USES
4.1 Alkylation.
4.2 Formation of long-chain alcohol monosulphates as surfactants.
4.3 Formation of intermediates in the preparation of some lower alcohols.
4.4 Minor uses are in the preparation of dyes, intermediates, stabilizers and speciality polymers.

5. DETERMINATION IN WORKSHOP ATMOSPHERE
5.1 Recommended Sampling Method
- Technique : Solid adsorber in collection tube
- Adsorbent : 70–80 mesh Poropak P in 7 cm × 4 mm tube – 150 mg in front portion separated from 50 mg in rear section by small polyurethane plug
- Sample flow : 0.01 to 0.2 l min^{-1}
- Sample volume : 0.25 l (at 1 ppm) to 12 l
- Extraction : Place both portions of Poropak P in separate 2 ml vials, add 1 ml diethyl ether, seal and stand for 30 min with occasional agitation

5.2 Recommended Measuring Method
- Technique : GC
- Column : 1.8 m × 3.2 mm stainless steel
- Support phase : 80–100 mesh Chromosorb WHP
- Liquid phase : 5% DEGS
- Temperature : Column 2 min at 50 °C then to 120 °C at 30 °C min^{-1} and hold; injector 180 °C
- Detector : Electrolyic conductivity detector in the sulphur mode (set according to maker's instructions)
- Carrier gas : He
- Carrier gas flow : 20 ml min^{-1}
- Sample size : 4 µl

5.3 Performance Characteristics
- Range studied : 1.8 to 24.5 mg m^{-3}
- Bias : (refer to 'Evaluation' section of ref. 1)

Dimethyl Sulphate

Overall precision : SD = 0.073
LOD : 0.25 µg per sample
Interferences : None identified

More detailed information is given in ref. 1.

6. OTHER METHODS

6.1 Various adsorbents, including Poropak Q (50–80 mesh) were used to trap DMS from 1 l air samples.[2] The adsorbent was contained in a 10 cm × 4 mm adsorber tube and the trapped DMS was removed from the absorbent by purging with helium with simultaneous heating. The DMS was flushed on to a 3 m glass column packed with 5% DEGS on Chromosorb W for GC separation prior to quadrupole mass spectrophotometric identification. DMS from air samples up to 20 l has also been trapped on silica gel prior to extraction with acetone and GC analysis.[3] This was performed on a 10 ft × 0.125 in stainless steel column packed with 10% OV-101 on Supelcoport (80–100 mesh) at 110 °C and using flame photometric detection. The limit of detection was 2 µg ml^{-1}. Fifteen litre air samples containing DMS were passed through a tube containing 100 mg silica gel.[4] Water was used to extract the DMS from the adsorbent and the extract was subjected to GC on a 3 m × 6 mm stainless steel column packed with 10% Apiezon L on Chromosorb P (60–80 mesh) at 100 °C. Detection was by flame ionization and the limit of detection was 0.05 ppm. In another silica trapping method prior to GC determination, diethyl ether was used as extractant.[5]

6.2 In the determination of DMS in workplace atmospheres, the DMS, having been desorbed from silica gel with triethylene glycol, was reacted with potassium cyanide to form methyl cyanide which was then analysed by GC.[6]

6.3 There has been one published method using high performance liquid chromatography (HPLC) to determine DMS.[7] Ten litres of air containing DMS were drawn through a NIOSH-type sampling arrangement of two tube sections containing silica gel and the DMS was derivatized *in situ* using *p*-nitrophenoxide in acetone. After the addition of sodium hydroxide solution and the internal standard, *p*-nitrophenetole in diethyl ether, the mixture was shaken and the ether layer separated for HPLC analysis. This was performed on a 25 cm Alltech C_{18} (10 µm) column using methanol–water (11:9) as mobile phase at 1 ml min^{-1} and using u.v. detection at 305 nm. The limit of detection is 1 µg. This type of method has been used to check the NIOSH one.

6.4 DMS and monomethyl hydrogen sulphate in air were trapped on a bed of XAD-II resin and were determined by direct probe temperature programmed mass spectrometry.[8]

6.5 A novel colorimetric method has been used to determine DMS in air after passage through pyridine.[9] The resulting solution of *N*-methylpyridinium methyl sulphate was reacted with hot sodium hydroxide solution to form glutaconic aldehyde which was reacted at pH 5.5 with aniline to form orange dianil. The absorbance was measured at 484 nm and the working range was 0.2–2.6 µg m^{-3} air. Spectrophotometry was also used after boiling DMS with quinoline to produce a coloured product[10] and in another spectrophotometric method, DMS adsorbed in sodium hydroxide solution produced methanol which was oxidized to formaldehyde, this being reacted with 4,5-dihydroxy-2,7-naphthalene disulphonic acid to produce a coloured solution capable of being quantified in respect of DMS.[11]

REFERENCES
1. National Institute for Occupational Safety & Health. 'Manual of Analytical Methods', 3rd ed., 'NIOSH Monitoring Methods', DHEW (NIOSH) Publication No. 2524, US Dept. of Health, Education & Welfare, 1984.
2. D. Ellgehausen, *Fresenius' Z. Anal. Chem.*, 1974, **274**, 284.
3. J. C. Gilland, Jr. and A. P. Bright, *Am. Ind. Hyg. Assoc. J.*, 1980, **41**, 459.
4. K. S. Sidju, *J. Chromatog.*, 1981, **206**, 381.
5. H. Blome and M. Hennig, Ger. Offen. DE3,400,134 (German Patent).
6. H. Frind and K. I. Trageser, *Fresenius' Z. Anal. Chem.*, 1987, **326**, 517.
7. R. G. Williams, *J. Chromatog.*, 1982, **245**, 351.
8. L. Hansen *et al.*, *Environ. Sci. Technol.*, 1986, **20**, 872.
9. D. Tomeczyk and J. Bajnska, *Chem. Anal. (Warsaw)*, 1973, **18**, 543.
10. S. A. Psaltyra *et al.*, *Gig. Sanit.*, 1970, **35**, 87.
11. W. Cao and H. Cong, *Zhongua Yutangyixue Zazli*, 1984, **18**, 290.

Hexamethylphosphoric Triamide

1. CA REGISTRY NO.
680-31-9

2. SYNONYMS
Phosphorictriamide-hexamethyl; Eastman Inhibitor HPT; ENT 50, 882; HEMPA; hexametapol; HMPA; hexamethylorthophosphoric triamide; hexamethylphosphoric acid triamide;
HMPT; HMPTA;
phosphoric acid hexamethyltriamide;
phosphoric tris(dimethylamide);
phosphoryl hexamethyltriamide;
tris(diamethylamino) phosphine oxide.

3. MANUFACTURE
By the reaction of $POCl_3$ with excess dimethylamine.

4. USES
4.1 As an aprotic solvent in organic chemistry.
4.2 As a solvent for spinning acrylic fibres.
4.3 As a chemosterilant.
4.4 As a polymerization catalyst.
4.5 As a selective solvent for gases.

5. DETERMINATION IN WORKSHOP ATMOSPHERE
5.1 Recommended Sampling Method
Note: No attention has been paid to the determination of HEMPA in the atmosphere, so the suggested sampling method would be subject to confirmation by experiment.

Technique	: Trapping (a) on a solid adsorber, or (b) in an impinger
Adsorber	: Solid – Tenax GC; liquid – ethanol
Sample flow	: 1 to 3 ml min^{-1} (suggested)
Sample volume	: 1 to 250 l dependent on concentration of HEMPA
Extraction	: For Tenax GC thermal desorption: if a liquid impinger is used it is recommended[1] that the solution for analysis contain a minimum of 200 mg HEMPA per 5 μl aliquot

5.2 Recommended Measuring Method

Technique	: GC
Column	: 45 cm × 4 mm i.d. glass
Support phase	: 80–100 mesh GasChrom Q
Liquid phase	: 1% Carbowax 20M
Temperature	: Column 100 °C; injector 170 °C; detector 140 °C

Detector	: Flame photometric at 526 nm; H_2 200 ml min^{-1} O_2 40 ml min^{-1}
Carrier gas	: N_2
Carrier gas flow	: 160 ml min^{-1}
Sample size	: 5 μl

5.3 Performance Characteristics

Range studied	: 0.1 to 200 ng per sample
Bias	: None quoted
Overall precision	: Not given
LOD	: 0.1 ng
Interferences	: None quoted

More detailed information is given in ref. 1.

6. OTHER METHODS

6.1 In a more recent GC analysis of chemosterilants, including HEMPA, the separation was carried out on a 50 cm × 4 mm i.d. glass column packed with 5% Dexsil 300 (or equivalent stationary phase) on GasChrom Q (80–100 mesh).[2] The column was temperature programmed from 100 °C at 6 °C min^{-1} until no more peaks appeared, detection being as in ref. 1.

6.2 Various chromatographic techniques were tried for the analysis of HEMPA, including TLC on silica gel.[3] However, GC was finally chosen using an 8 ft × 5 mm glass column packed with 6% diethylene glycol succinate on Diatoport S (80–100 mesh). The column was operated at 210 °C and a sodium thermionic detector was used.

Note: It is clear that if the determination of HEMPA in the workshop atmosphere is to be carried out, research into a suitable sampling technique is urgently required.

REFERENCES

1. M. C. Bowman and M. Beroza, *J. Ass. Off. Anal. Chem.*, 1966, **49**, 1046.
2. M. C. Bowman, *J. Chromatog. Sci.*, 1975, **13**, 307.
3. A. C. Terranova and C. H. Schmidt, *J. Econ. Entomol.*, 1967, **60**, 1659.

Arsenic Trioxide

1. CA REGISTRY NO.
1327-53-3

2. SYNONYMS
Arsenolite; claudetile; arsenious acid; arsenious sesquioxide.

3. MANUFACTURE
3.1 Prepared commercially by the roasting of arsenopyrites (FeAsS).
3.2 Occurs, in substantial amounts, in the flue dust during the smelting of Cu and Pb ores.

4. USES
4.1 Preparation of herbicides and defoliants.
4.2 As a decolorizer and refining agent in the glass industry.
4.3 As a wood preservative.
4.4 As a primary standard in quantitative analysis.

5. DETERMINATION IN WORKSHOP ATMOSPHERE
5.1 Recommended Sampling Method[1]
- Technique : Collection and concentration on filter
- Filter : 0.8 μm, Na_2CO_3 impregnated cellulose ester membrane filter
- Sample flow : 1 to 3 l min^{-1}
- Sample volume : 30 l (at 0.01 mg m^{-3}) to 1000 l
- Extraction :
  1. Transfer filter to clean beaker
  2. Add 15 ml HNO_3 and cover
  3. Heat, at 150 °C, on hot plate until sample volume is reduced to 1 ml approx.
  4. Rinse down cover and sides of beaker, and add 6 ml 30% H_2O_2
  5. Evaporate, on steam-bath, just to dryness
  6. Cool, add 10 ml 1000 μg ml^{-1} Ni^{2+} soln., mix for 30 min in ultrasonic bath, reserve solution for AAS analysis

5.2 Recommended Measuring Method
- Technique : AAS with graphite furnace
- Wavelength : 193.7 nm with background correction
- Injection : 25 μl; dry at 100 °C for 70 s
  Char at 1300 °C for 30 s, atomize at 2700 °C for 10 s.
- Aliquot : 25 μl

5.3 Performance Characteristics
- Range studied : 0.67 to 32 μg m^{-3} on 400 l sample
- Bias : Not significant

Precision	: SD = 0.075
LOD	: 0.06 µg per sample
Interferences	: Other particulates as compound

More detailed information is given in ref. 1; this method is suitable for particulate as well as As_2O_3 vapour. If particulate As only is required then the method in ref. 2 is applicable, but this is not suitable for As_2O_3 or AsH_3.

6. OTHER METHODS

6.1 Atmospheric particulate matter has been trapped on an organic membrane during an 8 h sampling.[3] The filter was digested in acetone, the solution was ashed, and the ash extracted with 50% nitric acid. This solution was taken for AAS analysis, measuring the As at 193.7 nm. The limit of detection was 0.1 µg ml^{-1} of solution.

6.2 Although the use of graphite furnace AAS for the determination of As has not been common, metals, including As, have been determined in flue dust from a steel works.[4]

6.3 A similar type of sample was acid-digested and the digest was analysed using an As electrodeless discharge lamp;[5] detection was at 193.7 nm and the limit of detection was 1.2 µg ml^{-1} of solution.

6.4 As_2O_3 was the subject of air pollution studies where air samples were passed through either 0.1N sodium hydroxide or 0.1N sodium chromate solution, this solution being analysed directly by AAS.[6]

6.5 In aerosols, particulate As_2O_3 was filtered from a 500 m^3 sample of air (during a 24 h period) through a Whatman 41 paper.[7] The paper was digested with nitric acid prior to using flameless AAS analysis incorporating an electrodeless discharge lamp with detection at 197.2 nm; the limit of detection was 0.1 ng.

6.6 In the determination of metals, including As, in particulates, various types of glass fibre filter were tested and the best was found to be Whatman EPM-1000.[8] Potassium chlorate–nitric acid was used as the digestion medium before evaporation to dryness and dissolving the residue in hydrochloric acid. EDTA was added before the AAS analysis using detection at 197.2 nm.

6.7 Recently, the arsine generation technique followed by AAS analysis has become the most popular method for the determination of As_2O_3 in air.[7-9]

6.8 Fibre-glass filters (10 in × 8 in) have been used to collect the particulate which was digested with sulphuric and nitric acids before measuring the As content by hydride generation after treatment of the digest with sodium borohydride as reducing agent.[10]

6.9 In measuring the As, Se and Sb content of air, samples were passed at 1.5–3 l min^{-1} through a membrane filter until 60 l had been sampled.[11] The filter was wet-ashed using nitric, sulphuric and perchloric acids and AAS was carried out on the diluted digest solution using hydride generation and detection at 193.7 nm; the limit of detection was 0.4 µg m^{-3} air.

6.10 Hydride generation and detection at 193.7 nm was also used to determine As in air after filtration through fibre-glass during a 24 h period followed by acid-digestion of the filter.[12]

6.11 X-Ray emission spectrometry has been used to determine elements, including As, in particulate matter from air.[13–17] Cellulose type filters were used to collect the particulate to give a thin film of not more than 6 μm thickness. Whatman 41 filters were generally preferred and discs were frequently cut from the filter to obtain replicate analyses.

6.12 Particulate has been filtered from air and As has been determined on the substrate by reaction with silver diethyldithiocarbamate and measurement of the resulting colour.[18,19]

6.13 A later spectrophotometric method was based on the reaction of As_2O_3 with sodium hydroxide and measuring the resulting absorbance at 222 nm.[20] A Whatman 41 filter paper was used to trap particulate at 3 l min^{-1} for 3 min before extracting the particulate with N sodium hydroxide solution. The method has a working range of 11–44 μg As_2O_3 per ml of solution.

6.14 Neutron activation analysis has been used to determine As_2O_3 but has found little support among analysts.[21,22]

REFERENCES

1. National Institute for Occupational Safety & Health, 'Manual of Analytical Methods', 3rd ed., 'NIOSH Monitoring Methods', DHEW (NIOSH) Publication No. 7901, US Dept. of Health, Education & Welfare, 1984.
2. National Institute for Occupational Safety & Health, 'Manual of Analytical Methods', 3rd ed., 'NIOSH Monitoring Methods', DHEW (NIOSH) Publication No. 7900, US Dept. of Health, Education & Welfare, 1984.
3. J. Y. Hwang, *Int. Clean Air Congr. 2nd*, 1970, 352.
4. J. M. Ottaway and D. C. Hough, *Proc. Anal. Div. Chem. Soc.*, 1975, **12**, 319.
5. E. R. Likaits *et al.*, *At. Absorpt. Newslet.*, 1979, **18**(2), 53.
6. R. B. Wells, Int. Cont. Air Pollut., Paper No. 9, 1979, 11pp.
7. H. Bernard and M. Pinta, *At. Spectrosc.*, 1982, **3**(1), 8.
8. R. E. Byrme, *Anal. Chim. Acta*, 1983, **151**, 187.
9. K. G. Brodie, *Am. Lab. (Fairfield, Conn.)* 1977, **9**(3), 73.
10. P. N. Vijan and J. R. Wood, *At. Absorpt. Newslet.*, 1974, **13**(2), 33.
11. T. J. Kneip, *Health Lab. Sci.*, 1977, **14**(1), 53.
12. J. Hubert *et al.*, *At. Spectrosc.*, 1980, **1**(4), 90.
13. J. Billiet *et al.*, *X-ray Spectrom.*, 1980, **9**, 206.
14. J. Wagman, Colloid Interface Sci. (Proc. Int. Conf.) 50th, 1976, Vol. 2, p. 171.
15. J. R. Rhodes *et al.*, *Air Qual. Instrum.*, 1974, **2**, 1.
16. J. R. Rhodes *et al.*, *Anal. Instrum.*, 1972, **10**, 143.
17. B. E. Ariz and M. Chessin, *Adv. X-ray Anal.*, 1974, **17**, 225.
18. C. Wachler, *Z. Ges. Hyg. Grenzgebiete*, 1967, **13**, 659.
19. E. C. Tabor *et al.*, *Health Lab. Sci.*, 1969, **6**(2), 57.
20. C. A. Snyder and D. A. Isola, *Anal. Chem.*, 1979, **51**, 1478.
21. R. Dams, *Meded. Farc. Landbouwwetensch., Rijksuniv. Gent.*, 1973, **38**, 1869.
22. J. F. Walling *et al.*, *Air Pollut. Control. Assoc.*, 1978, **28**, 1134.

Cadmium Chloride

1. CA REGISTRY NO.
7440-43-9 (Cd)

2. SYNONYMS
None

3. MANUFACTURE
Prepared by the reaction of hydrochloric acid with either the metal, carbonate, sulphide, oxide or hydroxide and evaporating to dryness to give the hydrated salt.

4. USES
Used mainly in photocopying, printing and dying, and in electroplating.

5. DETERMINATION IN WORKSHOP ATMOSPHERE
5.1 Recommended Sampling Method
    Technique   : Collection of particulates on membrane filter
    Sample   : 0.8 μm cellulose ester membrane filter 37 mm diam. in cassette filter holder
    Sample flow   : 1 to 3 l min^{-1} by personal sampling pump
    Sample volume   : 25 l (at 0.1 mg m^{-3}) to 1500 l
    Extraction   : 1. Transfer sample and blank to separate beakers; add 2 ml HNO$_3$ and heat at 140 °C (until volume is reduced to approx. 0.5 ml)
                       2. Repeat 1. twice more using 2 ml HNO$_3$ each time
                       3. Add 2 ml HCl and heat on hot plate at 400 °C until volume is reduced to approx. 0.5 ml
                       4. Repeat 3. twice more using 2 ml HCl each time
                       5. Cool and add 10 ml H$_2$O
                       6. Transfer solution quantitatively to 25 ml volumetric flask and bulk to volume; strength of final solution is 0.5N HCl

5.2 Recommended Measuring Method
    Technique   : AAS, flame
    Flame   : Air/acetylene, oxidizing
    Wavelength   : 228.8 nm
    Background corr. : D$_2$ continuum

5.3 Performance Characteristics
    Range studied   : 0.12 to 0.98 mg m^{-3} (25 l samples)
    Bias   : Not significant
    Overall precision: SD = 0.06
    LOD   : 0.05 μg per sample
    Interferences   : Background correction required to control flame absorption. Iron does not interfere at 20:1 ratio

More detailed information is given in ref. 1.

6. OTHER METHODS

Methods available for the determination of cadmium compounds in particulate matter divide into AAS and X-ray fluorescence spectroscopy (XRF).

6.1 Most AAS methods are similar to the NIOSH one. A fibre glass filter was used to filter particulate in a high volume sampling procedure and the filter was destroyed with 20% nitric acid plus potassium chlorate.[2] After evaporation to low bulk, hydrochloric acid was added and after dilution, the solution was analysed as for the NIOSH method. The limit of detection was given as 0.5 mg 100 ml^{-1}.

6.2 In a similar method, a glass filter (Whatman EPM-1000) was used and the analysis was as for the NIOSH method but with the addition of 0.4% EDTA before the AAS.[3]

6.3 In an examination of atmospheric air pollutants, including cadmium compounds, a cellulose-type filter (10 in × 8 in Whatman 41) was used for collection and analysis carried out by XRF.[4] A 24 h sample was taken at a low flow-rate and after collection, discs were cut and examined as 'thin films', using the cadmium Lα line and a flow proportional detection system with a PET crystal. The limit of detection was given as 2 ng cm^{-2}.

REFERENCES

1. National Institute for Occupational Safety & Health, 'Manual of Analytical Methods', 3rd ed., 'NIOSH Monitoring Methods', DHEW (NIOSH) Publication No. 7048, US Dept. of Health, Education & Welfare, 1984.
2. J. C. Hubert *et al.*, *At. Spectrosc.*, 1980, **1**, 90.
3. R. E. Byrne, *Anal. Chim. Acta*, 1983, **151**, 287.
4. J. Wagman, Colloid Interface Sci. (Proc. Int. Conf.) 50th, 1976, vol. 2, p. 171.

Chromates of Zinc, Calcium and Strontium

1. CA REGISTRY NO.
13550-65-9, 13765-19-0, 7789-06-2

2. SYNONYMS
Vary depending on the compound.

3. MANUFACTURE
3.1 Zinc chromate is prepared from a soluble zinc compound and a chromate, at pH 6.0 in aqueous solution.
3.2 Strontium chromate is formed by the reaction between a soluble chromate and a strontium salt.
3.3 A reaction similar to **3.2** is used for the manufacture of calcium chromate.

4. USES
4.1 Calcium chromate is used in the metal finishing industry, and as a depolarizer in batteries.
4.2 Strontium chromate is used as a corrosion inhibiting reagent, and, in this application, is preferred to zinc chromate.

5. DETERMINATION IN WORKSHOP ATMOSPHERE
5.1 Recommended Sampling Method

Samples : 5.0 μm PVC membrane filter 37 mm diam. (the BSWP PVE Millipore filters are not recommended as some chromate may be reduced to Cr^{III} before extraction)
Flow : 1 to 4 l min^{-1}
Sample volume : 8 to 400 l – duplicate samples taken

Analysis for Cr^{VI}

Sample preparation :
1. Transfer filter to 50 ml beaker
2. Add 5 ml of 2%NaOH–3%Na_2CO_3 solution
3. Purge headspace with N_2 to prevent oxidation of Cr^{III}, cover with watchglass and heat to near boiling point for 30–45 min; ensure volume is maintained throughout
4. Cool and transfer quantitatively with distilled water to a 25 ml volumetric flask, keep volume to 20 ml or less
5. Add 1.9 ml of 6N H_2SO_4, swirl to mix
6. Add 0.5 ml of a 0.25% solution of diphenyl carbazide in 1:1 acetone/water – this solution is used for analysis

5.2 Recommended Measuring Method
Technique : U.v. spectrophotometry
Wavelength : 540 mm
Calibration : Solutions of Cr^{VI} in 0.5N H_2SO_4

5.3 Performance Characteristics
Range studied : 0.5 to 10 µg m^{-3}
Bias : Not significant
Overall precision : SD = 0.084

Analysis for Zn, Ca, Sr if required

Sample preparation :
1. Transfer filter to 50 ml beaker with watchglass
2. Add 10 ml HNO, and 2 ml $HClO_4$
3. Heat to fumes of ClO_4^- and continue fuming for 5 min with cover in place; allow to cool
4. Add 10 ml water, warm to dissolve salts, and quantitatively transfer to a 25 ml volumetric flask, cool, and bulk to volume

5.4 Recommended Measurement Method
Technique : AAS
Flames : Sr, Ca; air–C_2H_2 reducing
Zn; air–C_2H_2 oxidizing
Wavelength : Ca 422.7 nm
Sr 421.5 nm
Zn 213.9 nm

5.5 Performance Characteristics
Range studied : Ca 2.6–10.2 mg m^{-3}
Zn not studied
Sr not studied
Bias : Ca not significant
Sr not significant
Zn not measured
Overall precision : Ca SD = 0.063
Zn not measured
Sr not measured

More detailed information is given in ref. 1.

6. OTHER METHODS

6.1 There is a semi-quantitative method available based on the same principle as the NIOSH one. This uses measuring tubes (Draegerwerk AG) containing dilute sulphuric acid, filter paper, granular quartz (0.5–0.8 mm) coated with 0.1% diphenylcarbazide solution and granular silica gel and quartz grains (0.5–0.8 mm) packed in the order of the air flow direction.[2] Chromate from air with diphenylcarbazide produces a violet colour in the silica gel, the intensity of which is proportional to the chromate concentration.

6.2 Cellulose filter absorbers have been used to trap chromate from air.[3] After extraction, the chromate was reacted with diphenylcarbazide, the method covering the range 1–10 μg Cr^{VI}.

6.3 PVC (0.8 μm) has been found to be a suitable medium for filtering chromate from the fume emanating from welding and allied processes.[4] Air was sampled in the breathing zone by a small pump at 1.8–2.2 l min^{-1} for 60 min and the PVC filter was extracted with 2% sodium hydroxide–3% sodium carbonate solution at a temperature just below the boiling point, before the reaction in acid solution with diphenylcarbazide to form the colour which is measured at 540 nm. The method is very similar to the NIOSH one.

6.4 If a non-destructive technique is required, there is a choice of X-ray photoelectron spectroscopy, nuclear magnetic resonance or scanning electron microscopy–energy dispersive X-ray analysis.[5]

6.5 The X-ray photoelectron spectroscopic method, being a surface analytical technique, can differentiate the species of chromium-containing compounds, whereas the electron microscopic method can yield morphological information. After collecting sufficient particulate containing chromate, it is possible to use an X-ray fluorescence analysis of a thin film (6 μm).[6]

6.6 In a novel method, after collecting chromate from industrial air samples, the catalytic effect of Cr^{VI} in the oxidation of o-dianisidine with hydrogen peroxide was exploited.[7] Dowex 50W-XB resin was used to remove interfering metal ions before carrying out the reaction which resulted in measuring a colour at 450 nm. The method is sensitive enough to measure Cr^{VI} in the range 0–30 μg with a limit of detection of 1 ng.

6.7 There is an amperometric method for the determination of chromate but it does not appear to be very specific.[8] Zinc, calcium and strontium can be conveniently measured by standard atomic absorption or emission spectroscopic techniques but in the presence of other substances containing the same elements this approach would be non-specific. Since it is the chromate ion which is the suspected carcinogenic moiety, methods based on chromate determination are preferred.

REFERENCES

1. National Institute for Occupational Safety & Health, 'Manual of Analytical Methods', 3rd ed., 'NIOSH Monitoring Methods', DHEW (NIOSH) Publication No. 7600, US Dept. of Health, Education & Welfare, 1984.
2. K. H. Huneke and W. Laufenberg, Ger. Offen. 2,913,283.
3. T. Dutkiewicz and J. Konezalik, *Bromatol. Chem. Toksykol.*, 1971, **4**, 293.
4. B.S. Method.
5. X. B. Cox *et al.*, *Environ. Sci. Technol.*, 1985, **19**, 345.
6. J. Billiet *et al.*, *X-ray Spectrom.*, 1980, **9**, 206.
7. B. M. Kneebone and H. Freiser, *Anal. Chem.*, 1975, **47**, 595.
8. H. Helbig, *Chem. Tech. (Leipzig)*, 1969, **21**, 553.